# Steelworker Alley

## HOW CLASS WORKS IN YOUNGSTOWN

*Robert Bruno*

ILR PRESS
an imprint of
CORNELL UNIVERSITY PRESS
*Ithaca and London*

**Library of Congress Cataloging-in-Publication Data**

Bruno, Robert, b. 1955
    Steelworker alley : how class works in Youngstown / Robert Bruno.
      p. cm.
    Includes bibliographical references and index.
    ISBN 0-8014-3439-4 (cloth : alk. paper). — ISBN 0-8014-8600-9
(paper : alk. paper)
    1. Iron and steel workers—Ohio—Youngstown.  2. Steel industry
and trade—Ohio—Youngstown.  3. Working class—Ohio—Youngstown.
4. Social classes—Ohio—Youngstown.  5. Class consciousness—Ohio—
Youngstown.  I. Title.
HD8039.I52U522  1999
305.5'62'0977139—dc21                       98-52184

Cornell University Press strives to use environmentally responsible suppliers
and materials to the fullest extent possible in the publishing of its books. Such
materials include vegetable-based, low-VOC inks, and acid-free papers that are
recycled, totally chlorine-free, or partly composed of nonwood fibers. Books that
bear the logo of the FSC (Forest Stewardship Council) use paper taken from forests
that have been inspected and certified as meeting the highest standards for environmental and social responsibility. For further information, visit our website at
www.cornellpress.cornell.edu.

Cloth printing      10 9 8 7 6 5 4 3 2 1
Paper printing     10 9 8 7 6 5 4 3

FSC FSC Trademark © 1996 Forest Stewardship Council A.C.
SW-COC-098

*For working-class people, like my parents, who never thought they were important enough, and for my daughter, Sarah, who needs to learn early in life that the past is always made, and never given.*

# Contents

# Acknowledgments

When I first thought about writing a book on working-class identity, I never fully anticipated the number of people who would play
both large and small roles in bringing the story to life. If the journey
began anywhere, it was with my wife's gift to me of Herbert Gutman's essays in *Power and Culture: Essays On The American Working Class*, edited by Ira Berlin. The book was the first scholarly text
that excited my interest in the power of working-class people. Gutman had passed away just a few years before I read the work, yet his
research approach and level of analysis revealed a truth about class
experiences that explained my own background. In addition to Gutman's work, the texts of Ira Katznelson and David Montgomery, as
well as Karl Marx's collective writings, provided my own research
with a framework for analysis.

There were also key individuals who at an early stage influenced
the direction of the work. Mark Roelofs encouraged me to think in
broad terms and not to fear a bit of ambiguity. Tim Mitchell constantly directed me to look into the many details of class life and to
find the nuggets that contradicted conventional wisdom. Last but
not least, Bertell Ollman was a good enough Marxist to challenge
each and every use of the term *class*. He was also a good enough
guide to support me in an exploration of what it meant to be working
class.

Along the way from idea to bound publication, I was fortunate to
meet people like Donna DiBlasio and Marian Kutlesa. Donna was at

the time the site manager of the Youngstown Historical Center of Industry and Labor. She opened the center's archives day after day and found room in the center for me to read and write. Marian is the warm and gracious director of the Struthers Historical Society. On my many trips back to Struthers, Marian interrupted her private life to open the society's facilities and allowed me to rummage for hours through historical documents. Without either of these women my book would have a great deal less history to characterize it.

Once words found their way to paper, scholars such as Michael Frisch were kind enough to review the work and offer valuable suggestions. No one read more of this manuscript than my editor, Fran Benson, at Cornell University Press. She asked a lot but always offered the possibility that something special would emerge. I'm grateful that she never hesitated to make criticisms and that she always did so with delicacy. Along with the efforts of the aforementioned people I'm also thankful for the support I received from my University of Illinois colleagues, Ron Peters and Helen Elkiss. Without their assistance, completing this book would have been impossible.

With the book now a reality I can admit the many times that the process could have been halted. Delays, additional drafts and requirements, short time frames and deadlines, and computer problems often left me dazed and confused. The fact that the work never faltered is a tribute to my wife, Lynn. She said, more often than anyone else, that I had a great story to tell. She was also the one who said that maybe others could help me say it more effectively. Lynn always reminded me to have patience and when necessary warded off my feelings of uncertainty. She proofread, criticized, encouraged, and fixed all of my crazy computer problems. As the final days of this work approached she also took our three-year-old daughter, Sarah, to play—somewhere else. What more could a working-class kid turned writer ask for?

ROBERT BRUNO

*Glen Ellyn, Illinois*

Steelworker Alley

# Introduction: His Silence Broken

Sometimes the past becomes meaningful only when it seems to be at risk. The meaning of my own history unfolded when it appeared threatened. Despite the enormous debt that I owe the past, I never truly understood its explanatory power until I had cause to worry about losing it. I had of course good reason to remember. Mine was a unified history with strong, formative role models and uncomplicated ways of viewing the world. The important elements are easily summed up. I grew up as the son of working-class parents with only the children of working adults as my friends. Most of my neighbors, like my father, worked in one of the three major steel mills in Youngstown, Ohio.

There was great freedom and security in this childhood. Front yards stretching up and down the block on Wilson Avenue in Struthers were magically transformed into athletic fields. No matter how many times my friends and I hit the television wires with an errant football or baseball, no one ever took a ball away. At night, wild games of kick-the-can and tag took place across backyards unobstructed by fences. The space was open and inviting. I was free to grab an apple from the Modarellis' tree or ride my bike in and out of the Henrys' driveway. There was no adult on the block who did not know my name or who would not buy a school raffle ticket. Actually, I wasn't known by my name. I was "Bob and Lena's kid." But for at least three decades of my life I never really understood what it meant to be their son.

By the time I graduated from Struthers High School in 1973, my father had worked for Republic Steel for over a quarter of a century. My grandfathers had retired a bit earlier in 1963 and 1965, but my uncle Frank continued a long stint repairing railroad cars for United States Steel (USS). Together, my father and grandfathers logged 115 years of mill service. Most area families with long work records had similar tenures in the mills or related businesses. The local mills employed approximately twenty-six thousand workers in basic steelmaking, and they were responsible for a good deal of the 18 percent of the country's total steel output originating from the area's dozen or more facilities.

I went off to college, taking advantage of the educational opportunity that union wages made possible. During my first summer back, thanks to Uncle Frank, I temporarily became one of the thousand "mill rats" employed at USS. Working as a fill-in on a "labor gang," I did every dirty, labor intensive-job there was. If a concrete block needed to be dislodged from beneath the open hearth furnace, I heard "Bruno, get a pick!" If the masons needed cement to repair the furnace's brick lining, it was "Bruno, get a bucket!" When the "checker-chambers" needed to be flushed out so that hot air could circulate up into the furnaces, someone yelled, "Bruno, get a drill!" Work was everywhere, and I could get as many hours as I wanted. Workdays often became work nights, and I averaged sixty or more hours a week. My back ached, and my lungs burned. The work was done in stifling heat. After just a few sixteen-hour sessions, I understood half of the workers' boast that steel mills were both the hottest and the coldest places on earth to work. This went on, without letting up, for about three months. What I could do for less than a summer, my father did for thirty-seven years.

I graduated from college and continued my studies in graduate school. I learned that I had been born in a period of unprecedented economic growth. Conventional sociological doctrine argued that during the 1950s America became a country burgeoning with war-deferred consumer appetites; and with national and international markets ripe for exploitation, every worker had hopes of becoming the proud owner of a middle-class lifestyle. It was during this period, according to some scholars, that the working class virtually disappeared.

The weight of post–World War II scholarship leaves little room for the existence of working-class culture. The influential work of such

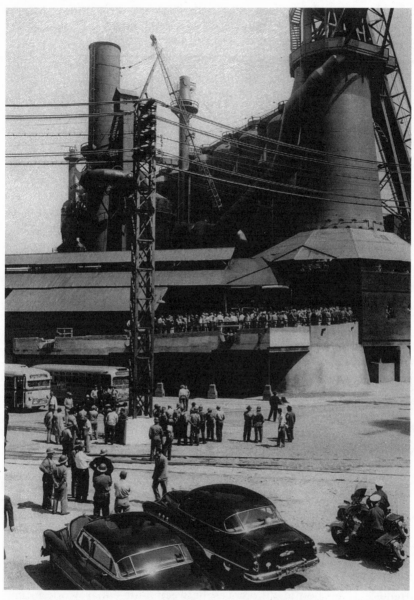

Re-lighting ceremony of USS's Ohio Works' blast furnace in 1951. Reprinted from *The Vindicator*. Youngstown, Ohio. © The Vindicator Printing Company 1998.

diverse intellectuals as Daniel Bell, Clark Kerr, Seymour Martin Lipset, John Kenneth Galbraith, Samuel Eliot Morison, Henry Steele Commager, Arthur Schlesinger Jr., and Peter Drucker applauded the seemingly self-evident truth that "everybody in America was middle class."[1] Even David McDonald, the president of the United Steelworkers of America, announced after a tour of local USS plants, "In America there is no class struggle."[2]

The social analyses I studied were based on the idea that workers thought of themselves as something other than members of a class. Depending on the approach of the scholar, workers might exhibit job, gender, race, or interest consciousness, but they were decidedly not conscious of class. Regardless of their theoretical stances, proponents of middle-class homogenization usually relied on some key factors to assess worker consciousness: (1) workers' affiliation with ethnic and religious identities that crossed class boundaries, (2) the absence of a labor party or revolutionary movement and, conversely, the presence of working-class support for a broad-based Democratic Party, (3) workers' relative rising economic prosperity after World War II, (4) increased occupational mobility and social stratification, and (5) ideological associations with middle-class habits and values.[3]

These denials of class consciousness shared a belief in the gradual and eventually broad assumption of middle-class values and behaviors by working-class people. Theorists like Bell and Lipset agreed in principle that as workers became more affluent they would reject traditional patterns of working-class sociability. The consequence of the workers' growing affluence would be a transformation of their social consciousness and conduct. With few exceptions, mainstream social science postulated that the long-term result of postwar capitalism would be the progressive deterioration of the working class. Theoretically, workers had rejected past beliefs and ways of living because they had become financially successful. Then, more or less automatically, they accommodated to middle-class norms and patterns of social relations.

The arguments for a middle-class identity, however, would have been more convincing if they had not overlooked at least five critical trends: (1) economic differences between blue-collar workers and managerial employees had not been sufficiently blurred by increased prosperity to erase class distinctions; (2) working-class lifestyles had not disappeared or even significantly eroded; (3) the workplace was still a site of conflict; (4) prosperity did not fracture working-class

communities built on family and friends; and (5) there were few un-
ambiguous signs that workers aspired to hold middle-class values.[4]

According to many social scientists and historians, I grew up in a
middle-class culture where people did not think of themselves as
workers, not even people who went off to steel mills everyday and
lived next door to others who did the same thing.[5] Somehow my father could be a steelworker and my mother a wage laborer, and most
of my friends the children of steelworkers, but incredibly, I had no
awareness of a working class that apparently trumped "a natural har-
mony of interests" unifying all Americans.[6]

I realize now that these notions of middle-class identity remained
ascendant because scholars paid insufficient attention to what workers actually did with their lives. With notable exceptions, such as
David Halle's *America's Working Man* and Michael Frisch and Milton Rogovin's *Portraits in Steel*, workers lives were studied not by
discovering how they lived but by myriad social indices. [7]

At the same time that capitalism was supposedly transforming the
class nature of American society, academics were selectively study-
ing, assessing, and statistically interpreting workers' opinions and organizations. These studies paid particularly close attention to workers'
"conceptions, images, attitudes, and ideational and verbal responses to
the social arrangements in which they [found] themselves."[8] Under
this framework, what passed for class identity was actually a snap-
shot of each worker's way of thinking.

But social theorists such as John Legget and labor historians such
as Herbert Gutman countered these ideas with the argument that
worker identity was more multifaceted than what could be revealed
by survey studies. Gutman noted that when class identity accounted
for the work patterns, social relations, religious practices, and everyday experiences shared by members of the community, it could be
more accurately defined as a product of social relationships.[9] These
relationships would undergo modification but at the same time remain surprisingly uninterrupted. According to this line of thinking,
my father's class identity should be understood in the ongoing context of friendships he had had for a lifetime. Leggett agreed, com-
menting that "[Too] often we treat class consciousness as a quality
either present or absent," when in reality "class orientation is sel-
dom ordered in such terms; rather it builds step-by-step (ordinal) as
on a continuum."[10] Class identity was developed in part through the
mutually reinforcing act of social recognition. In other words, it is

the totality of the "relations of class" which best defined worker identities. Joseph Schumpeter offered a fairly precise sense of how this transpired: "Class members behave toward one another in a fashion characteristically different from their conduct towards members of other classes. They are in closer association with one another; they understand one another better; they work more readily in concert; they close ranks and erect barriers against the outside; they look out into the same segment of the world, with the same eyes, from the same viewpoint, in the same direction."[11]

As if to sustain Schumpeter's insights, pre-1950 scholarship on labor and class bristles with phrases like "occupational neighborhoods," "plant-gate ghettos," "working-class communities," "class-based social places," "cultural patterns of behavior," "cultural rituals and religious practices," and "traditional modes of life." In these earlier analyses, there could be no confusion about where a student of the working class should go for understanding: the worker was an active participant in the formation of society and an agent of historical change.

Nevertheless, these views went out of fashion with the appearance of suburbia. Where workplace, neighborhood, community, and social relationships had once been dominant sites for working-class expression, they were now seen as modern forms of middle-class identity. If workers still mattered at all, indeed were even visible, they were shorn of their material lives. The point was too clear to miss: working-class identity was used up. Now nothing but less money separated steelworkers from steel executives.

I took in all these arguments about workers and their consciousness of class the way many students do, as engaging ideas. But they did not have immediate relevance to my life or to my own identity. After more than six years of higher education and a move away from home, I was little more than a respectful son, proud of my Italian surname. So I remained, until the past seemed about to disappear.

In 1987, I received a phone call from my father. I was living in New Jersey, several hundred miles away from my family. He explained that the doctors had found black spots on his lungs. He had just retired from thirty-seven years at Republic Steel, and now the doctors suspected cancer. More tests would be needed. He reassured me that there was nothing to worry about. My father's calm words were soothing but not very believable. His demeanor was always steady, and I could remember seeing him in an extreme emotional state only twice. When the casket was closed at his mother's wake, he collapsed across the coffin and cried in front of his wife and four boys. The second display was fury, not grief. What enraged this otherwise peaceful man was the realization that he had missed my first home run. When he wasn't working, my father usually accompanied my mother to the local ball field to watch me play baseball. But on this day—to be honest, I wasn't inspiring hall-of-fame dreams—he decided to stay home and paint. He was still on the ladder when we came home. My mother recounted the thrilling moment of my first four-bagger in vivid detail. Perhaps it was missing the sound of wood cracking against rawhide or the way the ball looked when it climbed into a blue summer sky, but Dad got really angry. He smashed the brush against the house and cursed loud enough to frighten the good fathers who lived two blocks away at St. Nicholas Church. A missed home run drove Dad a little crazy, but I could not recall him ever being worried about recurring layoffs, triennial strikes, too little money, or work injuries. Then, when he described his medical situation, it occurred to me that my father never worried about what he couldn't control.

His diagnosis was maddeningly equivocal. The doctor said that it could be cancer, but then added that it might not be. More tests were needed, and the progress of these black growths would just have to be monitored. The threat of my father's sickness loomed, and I realized that if he died now, he would be gone before I really knew him. Who was he really? Did he have dreams? How did he become a steelworker? What was his work like? Why were all his friends—all my parents' friends—from a place called the "mill"? Why did so many of my little baseball league teammates have fathers who worked in the same place? And why did everyone in my family love the taste of "beans and baloney"? The black spots had to be about something bad. I decided to go home.

Dad, Mom, and I talked. I asked them about matters no one else had ever asked them about. Dad is not given to many words, so it was not easy to get information. Mom, on the other hand, was not only spontaneously articulate but also embellished her responses in fine detail. In a few nights I had taped approximately four hours of oral interviews. Now I knew a few things. But still I felt unsatisfied. Dad and Mom spoke of experiences, places, friends, values, and relationships that seemed to flow out of some common stream. Their lives were about a way of work and a way of living that were shared with others who appeared to be very much like them. It occurred to me that my desire to know my father was more than an interest in personal history. He was a part of a neighborhood, an ethnic population, a community group, a union of workers, and he was a member of the working class. If he was all of these things, then, at least in terms of my formative years, I was, too. What began as a search for my father's identity turned into a search for my own. The past that we shared, which would explain so much of ourselves, was shared by thousands of steelworkers. So I bought more tape and stayed home a little while.

The steelworkers I came to know were the late-twentieth-century successors of Daniel and James Heaton, who, sometime between 1803 and 1805, fired up Ohio's first blast furnace on the banks of Struthers's Yellow Creek. In 1806, Turhand Kirtland, a local coal field operator, agreed to pay Dan Heaton $400 "for the bounty voted by the Company for the furnace, as he made the sufficient quantity of wears [sic] to entitle him . . . and what is of more consequence to the public, he has tried and proved the oar [sic] to be excellent."[12] The Heatons named the furnace at the Yellow Creek site in Struthers "Hopewell." The furnace burned kidney ore, wood, and charcoal to produce three tons of ore a week.[13] By the mid-1840s, iron furnaces were built in select locations to take advantage of rich coal and iron ore deposits. In 1869 at least sixteen local operations were generating bar iron worth an exuberant $168 a ton.

Just thirty years later, twelve area rolling mills were annually producing 150,000 finished tons for shipment to manufacturing centers in Pittsburgh. Shortly after the Civil War iron production gave way to steel manufacturing, and major capital investments were quickly poured into the construction of area puddling furnaces, muck bar mills, sheet mills, skelp mills, and tube mills. The area's first steel mills were built in the last two decades of the nineteenth century, and the valley's largest industrial employer, Youngstown Sheet and Tube (YS&T), was founded in 1900.

The company was capitalized out of $600,000 worth of common stock invested by six wealthy area families. The mill was built on an East Youngstown (later the city of Campbell) site purchased for $100 an acre.[14] Colonel George D. Wick headed an illustrious list of home-grown capitalists who quickly attracted the capital of other locally prominent "shrewd investors."[15] People whose names would dominate Youngstown's commercial and financial affairs for a century made sizable investments in the new steel giant. Soon the list of principal owners read like the city's social register; the Tod, Stambaugh, and Wick families accounted for up to twelve of the forty early stockholders.[16]

By the first decade of the new century, other area mills, like those at Republic Steel and United States Steel, were crafted out of million-dollar mergers of smaller steelmaking operations.[17] Numerous blast furnaces, open hearths, and rolling mills were developed out of Youngstown's rich iron ore foundries and constructed along the north and south sides of the Mahoning River. The cities of Youngstown, Struthers, Campbell, and Loweville then became home to the vast majority of steelworkers laboring in plants owned by the three national steel giants.[18]

With the outbreak of World War I, area steel production grew to support at least forty-eight open hearth furnaces and an eight-hundred-ton blast furnace.[19] Steel production was so significant that at the peak of America's World War II mobilization, Youngstown was nationally known as the country's "Ruhr Valley."[20] With the signing of a second armistice and a return to domestic work, forty thousand area residents made their living melting, pouring, and shaping steel. My father, both my grandfathers, one great-grandfather, three uncles, and two great-uncles were employed as steelworkers. All in all, about seventy-five thousand workers in the district took home a paycheck in related steel industries.

The primacy of steel in Youngstown continued until the morning of 19 September 1977. On that day, Youngstown Sheet and Tube announced it was closing down its mills and leaving Youngstown behind. The company's public statement, coming at the same time it handed written ones to union officials, was short on details but clear in its effect: "Youngstown Sheet and Tube Company, a subsidiary of Lykes Corporation, announced today that it is implementing steps immediately to concentrate a major portion of its steel production at the Indian Harbor Works near Chicago. . . . The Company now employs 22,000 people. The production cut-back at the Campbell Works will require the lay-off or termination of approximately 5,000 employees in the Youngstown area."[21] Workers and residents alike called it "Black Monday," and most still know the day by that name.

Youngstown Sheet and Tube's withdrawal was followed by the exodus of the industry giant United States Steel in 1979 and 1980 and the merger, and subsequent bankruptcy in the mid-1980s, of Republic Steel.[22] With almost cruel consistency, each fall from 1977 to 1979 renewed the destruction of the area's industry and heightened the insecurity of steelworking families. At the end of the decade ten thousand steelworkers had lost their jobs, and out of a post–World War II production workforce exceeding forty thousand, less than a thousand remained active as the 1992 presidential election drew near.[23]

The U.S. steel industry did not collapse just in Youngstown. In 1977 alone, 21,940 permanent jobs had been lost and nineteen plants closed in eight states.[24] The devastation was so complete that by 1982 monthly national employment in basic steel production had dropped by more than 50 percent (232,600) from its post–World War II high.[25] Steel production had fallen 20 percent by the middle of the 1980s, and unlike other slack times, this drop represented a permanent loss in capacity. During the recession of 1982–83 the nation's steel mills were functioning at only 43.8 percent of capacity.[26] But that did not mean that steel companies were doing badly. Throughout the 1970s the industry earned over "$1 billion a year in after-tax profits."[27] As U.S. Steel chairman David Roderick blithely commented, "U.S. Steel is not in the business of making steel. It is in the business of making money."[28] This is best illustrated by U.S. Steel's decision to change its name to USX in order to reflect its contemporary focus. In other words, the steel companies were gradually moving away from employing steelworkers.[29]

By 1987 there were very few active workers left, but the city was teeming with "retirees." As I waited for the results of first one, than many more of my father's medical evaluations, I interviewed seventy-five of those retired steelworkers who were still living in the Youngstown area. Each of these workers was employed by area steel manufacturers (see appendix, roster) by the early post–World War II period, and all of them participated in either the 1937 Little Steel Strike or the industry's longest strike in 1959. Since retiring, these workers have chosen to remain residents of the Youngstown metropolitan area (see map of Youngstown area). Thus, 96 percent of the workers interviewed were drawn from the primary manufacturing cities hosting steel companies, including Youngstown (37 percent), Struthers (37 percent), and Campbell (22 percent).

Steelworkers in Youngstown were predominantly foreign-born Italian, Croatian, and Slovakian immigrants or the children of

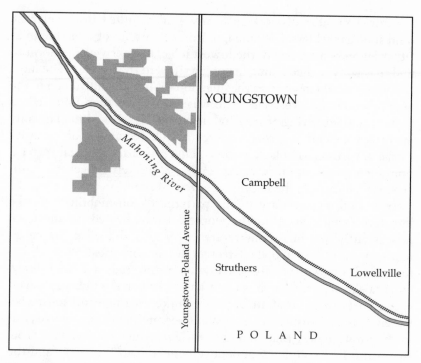

Map of Youngstown area, Mahoning County

foreign-born ethnics.[30] By 1880, approximately 90 percent of all industrial workers in the area were foreign born or the children of immigrants.[31] Consequently, 76 percent of those workers interviewed were of Italian or non-Anglo-European descent. While exact figures on minority employment in each mill are hard to find, precise records from Youngstown Sheet and Tube's Campbell Works reveal that in the early 1970s, 23 percent of their production workforce was black or Hispanic. My study has a 23 percent minority-worker representation.

Workers in this study represent the full hierarchy of job classifications and every phase in the making of steel (refer to appendix, interviews). A note of caution, however, is in order. The overall appearance of the job titles in this study seems to suggest a tilt towards skilled trades and perhaps represents the views of a more privileged laboring class. The appearance is deceptive in the absence of a knowledge of the job progression available to most long-term mill workers.

While a number of workers in my study worked in the electric weld shop as welders, tally men, or millwrights, all of them began as either laborers or in one of the lowest job classes. It was the norm to find that workers had a long list of job titles to their credit. The most common initial response to the question "What job did you do in the mill?" was "I did every job there was." This was then followed up with a detailed listing of jobs and departments worked in. The only important exceptions to this were the job sequences of minority workers, particularly black workers. Due primarily to racial segregation, the listings for black workers more accurately reflect their mill experiences than do the listings of white workers.

An important corollary to the patterns of job mobility was the long-term tenure of all of the workers I interviewed. Youngstown was essentially a one industry area. Not only did most people directly owe their livelihood to steel production but most people did so for their entire working lives. Workers in this study averaged thirty-one years of service, which is remarkably close to the industry's average of thirty-three years. Most of the workers (61 percent) started between 1946 and 1951, with 23 percent beginning in 1936 and 1937. It is important to note that all of the workers I studied left the mills between the first shutdown announcement of YS&T in 1977 and the first bankruptcy announcement of Sharon Steel in 1990.[32] These active working years constitute the period on which this book is based.

In this study I made an intentional decision to address the experiences of average workers. By average worker I mean someone who, in the words of most of my subjects, "goes to work every day, does [his] job, and collects a paycheck." I purposely focused on workers who were not likely to be the most obviously class conscious. This meant not relying on union officials or known "radicals." With a few notable exceptions, the workers in this study held no elected or appointed union position, and they were no more than irregular attendees of union meetings. Because this study is less focused on institutional behavior than it is on informal group actions, the practices and opinions of union leaders or political activists are not given special priority.

My pool of possible workers came from three sources: (1) District 27 Steelworkers Organization of Active Retirees (hereafter referred to as SOAR), (2) steelworkers mobilized into the Youngstown Chapter of Solidarity USA, and (3) referrals from individual workers. A few words of explanation about each of these sources are in order.

Some Youngstown retirees have chosen to have their continued economic and political interests represented by one of two organized groups. The larger, more financially secure group is the district chapter of SOAR. SOAR is actively supported by the leadership of the United Steelworkers of America, and the two promote identical agendas. SOAR members in Youngstown pay a dollar a month in dues and attend monthly meetings chaired by representatives of the old union leadership of Local 1331.[33] My father is a SOAR member and took me to a couple of local chapter meetings. At one of these sessions I was permitted to address the body- at-large about my work and ask for volunteers to participate.

The alternative, a more publicized and smaller organization is Solidarity USA. Unlike SOAR, Solidarity USA is not exclusively a steelworkers' retiree organization. While it grew out of steelworkers' resistance to mill closings and company bankruptcies, its members include working-class people from different local industries. Solidarity members do not pay union dues, and the organization has a more informal structure. Solidarity usually promotes economic and political positions independent of the United Steelworkers of America (USWA) and consequently is not formally recognized by the International Union.[34] Out of the seventy-five oral interviews I conducted, twenty-four of them were with workers who either attended SOAR or Solidarity meetings.

That still leaves 70 percent of my subjects as unaffiliated workers. In examining the attendance rosters of both groups and talking with a sample of active members, I found that most Youngstown retirees are not organizationally linked to the labor movement. In gaining access to these "independent" workers I was guided by one simple rule: let the worker you speak to open doors to the next worker. I recruited other subjects from the referrals, leads, and suggestions given by the previous worker interviewed. These referrals were often to a "friend," but many times they directed me to a small, informal gathering of retirees.

It is interesting to note that retired and displaced Youngstown steelworkers have not withered away or withdrawn into isolation. The vast majority of retired steelworkers (82 percent) did not move out of the area when the industry capped its postwar retrenchment with wholesale shutdowns in the late 1970s.[35] Steelworkers continue to live in the valley, and many have maintained their loyalties toward one another. I found groups of retirees meeting on a monthly basis for breakfast and lunch all over the county. A group of seven or eight workers from Local 1331 meet every Wednesday for lunch. Another 1331 group gets together on the first Wednesday of every month. Sharon Steel employees congregate once a month at the Country Kitchen. Workers from YS&T gather around a conference-room-sized table the first week of the month to break bread with each other.

In addition to these more intimate associations, most of the local unions have active retiree organizations, even though all but one of the unions are defunct. I was fortunate enough to be a welcome guest at the YS&T Local 1418 retirees' monthly social. Workers, their wives, and the spouses of deceased local members regularly planned dances, picnics, and fund raisers.

I conducted both group and single discussion sessions with workers. Group sessions, involving 21 percent of the workers, usually included two but not more than four participants. I entered every session with thirty questions (see appendix, discussion questions) derived either from workplace or community activities. While I always worked from the same framework of questions, some questions could break down into derivative inquiries. For example, the question "Did you know any bad union members?" could extend to "What did they do?" depending on the first answer. Given, however, that the majority of questions (26 out of 30) addressed worker activity, most of them could splinter into "if . . . then" types.

In addition to speaking to workers, I invited and encouraged their wives to take part in our conversations. While most sessions did not include husband and wife, a quarter of the workers interviewed had a spouse present. Since most of my discussions were held in the homes of workers, there was, at times, the opportunity to involve a spouse during the interview. In these situations, the presence of a spouse served two important functions. First, the wives of workers often remembered in finer detail the experiences of community life. This was not surprising considering that only 3 percent of them worked while their husbands were employed and, consequently, engaged more hours of the day within the home. What was surprising, however, was that in only a few cases did wives express a greater knowledge of people and events within the community than did their husbands. Workers and their wives had roughly equal knowledge of the same social experiences, but what they stressed about each person, place, or event was sometimes different.

Wives made a second important contribution by providing information that elicited either an additional worker response or that stimulated a new direction or question for consideration. For example, one woman mentioned how some men used to work during their vacations. I had not previously asked nor planned to ask her husband about such a practice. His answer, however, and the discussion that ensued about working during vacation time contributed to understanding an important area of worker discontent. In other cases, workers were just stuck for an answer or unable to provide an example of something they had said. Either they couldn't remember or, according to their wives, remembered incorrectly. On these occasions, I was fortunate to have a spouse at the table who could detail three or four examples of experiences with steelworking friends.

While the exchange of class memories is important to a rich personal accounting of what it meant to be a steelworker in Cold War America, my study did not rely exclusively on oral histories. In support of the oral testimony, I conducted a social historical analysis focusing primarily on one of the Steel Valley's typical steel towns, the city of Struthers. Struthers was home to two thousand steelworkers in 1952 and to 33 percent of the workers I interviewed.

Throughout the 1950s and 1960s steelworkers represented over one third of the city's population. This was the so-called "end of ideology" period. My interviews with the workers focus on the post–World War II years through 1977, a time generally considered relatively bereft of expressions of class identity and resistance. This, however, was not true of the steel industry. After short strikes in 1950, 1952, and 1956, the United Steel Workers of America conducted and subsequently won in 1959 the longest strike in the history of the steel industry. Those years were also a time of unreported shop-floor "wildcats" and in-house strategies of worker opposition.

Struthers was also an exception to another national trend. Cold War political, economic, and cultural influences, as well as assaults on organized labor, have been viewed as the final nails in the coffin of most variants of working-class politics.[36] Yet, despite national voting patterns and two-party entrenchment, events in Struthers made it clear that on a local level steelworkers actually made the laws.

While there are many competing contemporary models for depicting class, the one that best describes the experience of the Youngstown area steelworkers is formally presented by Ira Katznelson. Katznelson, unlike many other commentators, explains that working-class identity occurs within four layers of social development: economic structure, ways of life, worker dispositions, and collective action.[37] *Economic structure* refers to the principal characteristics of a profit-driven economy where workers are paid a wage below the value of their labor. On this level, workers know they are working class because in order to survive they are dependent upon receiving wages for the work that they do.

The economic structure of society is a precondition for class *ways of life*. At this level, workers reflect a class identity in how they live and interact both within the workplace and the community. Ways of life are built upon long-standing, formative experiences growing out of intense social relationships. Class is not viewed one-dimensionally as the product of having to work for a living, but is given meaning in the social interactions of workers at home, as well as on the job.

Workers' ways of life can further promote the same universe of "interests, social experiences, traditions and value-system[s]" among one another. Class identity is exemplified by the *disposition* of workers. According to Katznelson these dispositions are "formed by the manner in which people interact with each other" and are not merely individual opinions. Shared perspectives and common vantage points will, upon sufficient provocation, then lead workers to *collective action*. When they do so, they act consciously in pursuit of common goals. At this level workers act decisively on behalf of their common condition and needs.

Katznelson proffers his theory of class formation as a way to better understand the complexities of class as an analytical concept and to avoid the often reductionist quality of mainstream work. But he does not systematically apply it to any particular American working-class population. My work, therefore, can be seen as a modest empirical application of Katznelson's thesis to a classical representative case: a postwar, industrial steel town.

Besides a small mill home and well-worn used cars, my parents owned no capital, no business, no machines, no rented property, no labor other than their own, and no investment money. While I cannot remember growing up with any conscious sense of the word "class," I also could never connect what I later knew about middle-class prosperity and values with the way my parents lived. The dissonance I felt between accepted doctrine and my own experiences motivated me to look for meaning in my past by examining what postwar commentators appeared to have written off or at least had come to comfortably define—the working class. Despite the scholarship that discarded class as a historical concept, I became more and more certain that I was raised from cradle to college in a working-class family. That sense of class is at the heart of this book.

The accounts that the steelworkers shared with me concerning their working lives strongly contradicted the idea of a lost working-class identity. Far from being a class-dulled industrial labor force, these workers lived according to class. Class informed every facet of their lives and their life stories. When they spoke of their friends, they cited other workers. When asked about their economic conditions, they noted their meager earnings and irregular work. On the many occasions I asked about socializing, they took me to clubs, picnics, and parks where there were other workers. Neighbors who also happened to earn their living in the mills shared their unfenced yards, gardens they tended, homes they painted, roofs they attached, and lawnmowers they repaired.

My questions about their post-war employers elicited comments sounding remarkably like the radical language used by their allegedly more class conscious Depression Era brothers and sisters. A them-versus-us mentality had flared up in big and small ways on the shop floor, and when they voted locally, it was always for the "friend of the working class." What all this indicates to me is that workers from the early fifties to the late seventies recognized their class identity in at least four ways: (1) by the way in which they lived among each other, (2) according to their common dispositions and attitudes, (3) in the material condition of their economic life, (4) and in the forms collective action and resistance they took. They paid attention to how workers acted in and out of the mill, and they gave priority to doing things with other workers. In my interviews their class consciousness emerged as a way of life.

My father did not get cancer, and, as this book goes to press, he continues to live a robust life. I initially chose to examine the lives of Youngstown steelworkers because my father was one. But on further reflection, I became keenly aware that worker ideas and actions are always intertwined with a particular time and rooted in a familiar, known place. While there are remarkable similarities across locations, which make it possible to talk about a national working class, the class identity of Bethlehem, Pittsburgh, Gary, Buffalo, and Chicago steelworkers necessarily includes their particular community structure. Within each of these local histories lies a treasury of stories that bridge the personal with the universal, the worker with the class, the resident with the community, and the person with the family.

# 1  Steel-Paved Streets

The first time I came back to the Youngstown area, the place looked much the same as it had before I permanently relocated in the late 1980s. In downtown Struthers, the Wagon Wheel was still serving cheeseburgers, the Center Street bridge still needed serious repair, and the Silver Mirror, at the corner of Midlothian and Youngstown-Poland Roads, still offered the best breakfast deal in town. Driving through old neighborhoods, I realized there were fewer trees lining the street and cars in the driveways were older. But the houses were exactly as I remembered them. Most had front porches, clean siding, and small, well-groomed yards. Despite the collapse of the steel industry, Struthers had remained remarkably healthy. This was due in large measure to the pension funds of thousands of retired union steelworkers.

Struthers was a community where notions of class were formed in part by residential housing patterns and the ways of life that grew out of these spatial arrangements. The workers I interviewed described a classic industrial community during America's greatest period of economic growth, and although the 1960s brought to the Youngstown area some of the same suburbanization and exodus from the center of the city that occurred throughout the rest of the country, these changes did not dismantle the working-class communities where the steelworkers lived.

While the workplace unquestionably reinforced class identity, most postwar Youngstown steelworkers already had a class orientation

when they entered the mills. Attitudes that came into play on the
shop floor were often formed in and reimposed by the neighborhoods
that workers established.[1] Beginning before World War II and continuing well into the 1970s, steelworkers lived near other steelworkers.
For long, uninterrupted periods workers' lives took place within arm's
reach of each other. Limited space in the neighborhoods and close
proximity to the mills created an interdependent relationship between home and work. The physical dimensions of industrial towns,
the density of working-class residences, the openness of neighborhood
streets and yards, and the monolithic presence of the steel industry
were the critical variables that shaped the workers' social interactions.

If there was anything workers respected more or held in higher regard than the workplace it was the place they lived. Workers knew
with a certainty equal to their knowledge of how their jobs were to
be done the intricate contours of their neighborhoods. In interview
after interview, workers stressed that they not only toiled alongside
one another but spent most of their nonworking time close to a "mill
hand." The influence of common living spaces on class formation
was not restricted to the shop floor.

## Where Workers Lived

Youngstown area steel mills dominated the terrain along the Mahoning River Valley (see map of Youngstown area steel mills) in
northeastern Ohio.[2] Area steel companies were dominant employers
and consumers of materials and supplies. The concentration of steel-
related production in the Youngstown district was proudly thought
to be "the greatest on earth."[3] Steelmaking's tremendous demand for
primary and secondary goods spun off auxiliary manufacturing industries. Basic steel and related production had grown to such an extent
that even during the downsizing of the mid-1970s, approximately
eighty thousand people were employed in manufacturing, accounting
for a $1,108,119,000 payroll.[4]

Struthers, as well as Campbell and Loweville, was constricted in
its pattern of residential development and proximity to the point of
production. In the city, Youngstown Sheet and Tube (YS&T) constructed mill homes along both sides of the Mahoning River. Shortly

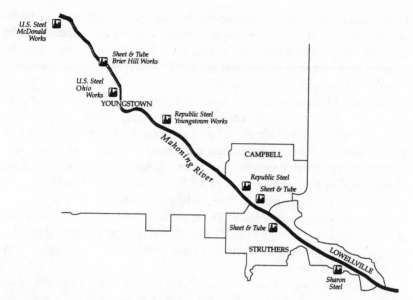

Map of Youngstown area steel mills

after incorporating in 1900, the company began to purchase land from large private estates through its real estate development subsidiary, the Buckeye Land Company. By 1913 it had already plotted a considerable amount of land, and following the 1916 "steel riot" in East Youngstown it allocated $250,000 for the construction of workers' homes.[5] Eventually YS&T land holdings stretched from the river's edge to cover over half the territory Struthers had dedicated before 1930.[6]

Thus, within roughly one square mile, successive generations of working-class families established residence in company housing. By the late 1960s, home ownership was widespread, although the town newspaper's wedding announcements occasionally revealed that a newlywed couple would reside "with the husband's parents." Buying a house meant a degree of prosperity. Losing one was a reminder of how tenuous that prosperity could be. In 1959 the United Steelworkers of America (USWA) waged a 116-day strike. It was not the bloody, confrontational strike of the 1930s, but it still bore serious costs. My family was forced to sell our first home and move back in with my father's parents. I may have hated leaving that house because its

cellar was large enough to serve as a hippodrome for me on my tricy-
cle, but losing their own home was traumatic for my parents. Mom
said she never liked the idea of going back to her in-laws'. All she
wanted was "a place for us, our family." "What else," she wondered
aloud, "was working hard about?"

Most of the homes were small, two-story, box-framed structures
facing the curb with tiny yards. Long porches, perfect for visiting,
were set neatly alongside gravel and cement driveways. Certain city
streets were so heavily populated with steelworkers that they could
have been credibly renamed "Steelworker Alley." The best example
of steelworker concentration was Wilhelm Street on the north side of
town. Wilhelm ran through the part of town rising up on the east
side of the Mahoning Valley. From their doorsteps residents could
see the broad rooftops of YS&T's Rod and Wire and Campbell Works.
On this street alone thirty out of a total of forty-two residences were
owned by mill hands.[7]

Another such street was Sexton. Out of 458 homes running east
and west for nearly half a mile, one quarter of them in 1952 were occupied by steelworkers. Other streets had steelworker densities of 15
to 20 percent from 1950 through the 1970s.[8] Now it is important to
realize that I have calculated only the number of people who worked
in one of the area mills. If I added the number of people who worked
in related manufacturing fields and those who held manual labor
jobs, the percentage of working-class people living side by side would
have likely reached 95 percent on a majority of streets.[9]

Since most workers were employed in the steel industry, they had
close knowledge of many other workers. When asked who their closest friends were, workers never failed to give the same answer—other
workers. Friendships are made when people inhabit the same space.
"They were all from the mason department," the workers told me.
"They lived all over the street." "A lot of the guys I worked with I
also grew up with." "My best friends were from Puerto Rico and also
worked in the YS&T." "I met them here in the mill." "They were
guys I met in different departments, and in the golf league."[10]

It is not surprising that workers claimed to have only steelworkers
as close friends. Except for other manufacturing-based jobs, there
were few other occupations represented in the social spaces shared
by the steelworkers. It has been estimated that during World War II the
Youngstown area had sixty-five thousand people employed in steel or

The street where my father grew up. Notice the working-class homes snuggled together.

iron production, and throughout the postwar years there were approximately fifty thousand members of District 26 of the United Steelworkers of America.[11]

Steelworkers were not just work buddies. They were neighbors, and class formation was a product of neighborhoods as well as eight-hour shifts. As YS&T employee Brad Ramsbottom pointed out, "Steelworkers lived on either side of me and one guy lived across the street. We all had kids about the same age, and seven of us worked at YS&T. It was a working-class neighborhood." My uncle Frank lived within ninety feet of my grandfather and less than a block from my great-grand-uncle Louie. Each of them lived next to other steelworkers. They may have worked for different mills, but they lived on the same street. The neighborhoods were critical to social interaction, because friendships were not usually bound to the workday. Consider the way my father answered the question "Were your closest friends steelworkers?" "Yeah. Kenny Wanamaker. We went to the lake together. His in-laws had a small cottage on Lake Erie. We bowled together too. Bill Mastrangelo worked there [the mill] for a short time. Most of these guys I knew from outside of the mill. I knew them from around the area and in Struthers. We socialized a lot there. We went to the youth center where the present city hall is now. I met Louie Veltrie there and he worked for Republic."

Many of my father's best friends were steelworkers whom he knew before entering the mill. This history was common for all of the workers interviewed. An analysis of the 1940 graduating class at Struthers High School revealed that nearly 60 percent of the men whose occupations could be identified were employed by area mills in 1952.[12] The workers identified at least half of their closest work buddies as men they had known from their high school days. In fact, two of the friends that my father referred to, Bill Mastrangelo and Louie Veltrie, were in the same graduating class.

Workers spent most of their lives doing things with and for other workers in the community. It was not uncommon for one worker to do small household jobs for another. Some workers, like Jim Visingardi, even did skilled manual labor. When asked if he ever helped other workers outside of the mill, Visingardi quickly exclaimed, "Oh yeah. I would put foundations in for a garage or a little concrete slab in the back of the house. Small jobs really. Half a day of work." He then proudly added, "You never charged anybody."

"My house was built by guys from the mill," John Zumrick told me. "One knew roofing; another, something else. We had a big party here when my house was finished, and my wife made all the guys lunches and we went right to work. It was like a big family." In Zumrick's case, the distinction between home and work was blurred. The guys he worked with built his home. The locus of his private life— his house—was a material representation of collective labor, and it was built between work shifts.

Working days were wrapped around collective activity outside the plant gate. Zumrick and his friends went directly from home building to steelmaking. They never stepped clearly away from the familiar realm of home and intimacy into a space dedicated purely to economic ends. What the workers experienced inside of the plant came home with them, and what they did in the community was inextricably linked to their mill relationships.

When workers interacted cooperatively outside of the plant gate, it was undoubtedly because they felt a strong connection with each other. That connection was as much literal as it was metaphysical. In the city of Struthers, a classic blue-collar community, steelworkers accounted for roughly 26 percent of the town's total mill-age-eligible adult male population, and three-quarters of them lived within one square mile of YS&T's Stop 14 main entrance.[13]

My family was no exception. In 1959 we moved out of my father's parents' home, which sat at the top of a hill in Youngstown, just up the block from where Dad worked at Republic Steel. The home we purchased on Wilson Street in Struthers was now a bit further from Dad's Republic Stop 5 entrance, but it was actually closer to where his father worked. My grandfather was employed by YS&T, and he used the Stop 14 gate on State Street in Struthers to enter the plant. The move slightly complicated their routine of riding to work together, which they had begun while they were still living in the same house. My grandfather usually showered and changed after work in the mill, and he often came out of the plant later than many of the others. When my dad was not around with his used 1937 Chrysler, his father caught the bus home. But after those 4 P.M. to 11 P.M. shifts when his father was a little slower than normal, he would miss the last bus back to Youngstown. Then, according to my father, "he had to hoof it home."

## Deep Roots

The spaces the workers inhabited were well-traveled, contiguous, and practically permanent locations. Their stays in steel towns like Youngstown, Loweville, Struthers, and Campbell were often for a lifetime. Most were locally born and bought their first homes close to their parents'. Even home ownership was an act of cohesion. A large number of workers bought homes between 1946 and 1959 and settled in a large area covering 205 acres originally owned by YS&T's Buckeye Land Company. The land would later become part of three communities, numerically identified as tract numbers 9905, 9907, 9895, and 9898. YS&T gained ownership of this land through probate court and immediately developed housing for workers and foremen. In 1949, Mike Vasilchek purchased property within tract 9907 for $5,800.[14] His home (lot no. 44647) was built in a cluster of "bosses' houses," and it was less than a half mile from the entrance to YS&T's Stop 14.

YS&T Stop 14 Entrance, State Street and Walnut in Struthers. My father's dad and thousands of other workers walked across the bridge to start and end the work day. Reprinted from *The Vindicator*. Youngtown, Ohio. © The Vindicator Printing Company 1998.

At 52 by 128 feet, however, Vasilchek's lot was considerably bigger then the average mill-workers' property. Mill homes were usually situated on lots that averaged 45 by 100 feet. But many lots in Struthers measured a more modest 35 by 75 feet, and those located within a stone's throw from the river were smaller yet. George Alexoff's home on Prospect Street was built on a 50-by-49 foot lot that was approximately 34 yards from the mill. Housing lots on Bridge, Terrace, Main, Liberty, Hazel, Moore, Reed, Grant, and Union were within 200 feet of YS&T's Rod and Wire and Campbell Works. Some of these "plant gate homes" were built on tiny 18-by-66, 66-by-66 and 38-by-80 foot lots.

The property development that extended further out from the river's banks remained within 300 yards of Stop 14, [15] and land development outside the mill's shadow did not reduce housing density. Some workers, like Joe Flora, lived in homes a little farther away, but still within a short walk to the plant. Flora is a very easygoing man, who had recovered from a brain tumor a few years before we talked. Throughout our conversation we recalled the years that he and his wife, Eileen, knew my family, and how often as a kid, I had played with their son (Bob). It was easy and fun going to Bob's house because at that time the Floras lived just one street away, and it was very near a short-cut entrance to the park. As we spoke in Flora's kitchen under a whirling ceiling fan, he recalled how he had come to own that house near the "path" to the park entrance.

In 1957, Henry Higgins, Flora's father-in-law, was on his way to work as a load dispatcher in the powerhouse at YS&T when he died of a heart attack. Flora then acquired Higgins's home on Creed Street, from which he followed the same path to work his father-in-law had taken. Flora's second home, purchased in the early 1970s, was located further from the mill. While the home was bigger and afforded him an electrical shop in the basement, it was a good walking distance from the first. My father purchased a home in Struthers in 1959 and twenty years later moved exactly one street away.

Jim and Josephine DeChellis bought one of the first homes constructed on Wilson Street between Fifth Avenue and Eighth Street. Their house is less then fifty yards from Joe Flora's first home on Creed, and the two would sometimes walk to work together. The DeChellises had lived in the same house for forty-four years, and our conversation took place in the same kitchen where Josephine has always poured coffee and served cinnamon rolls for visitors. Jim explained the concentration of steelworkers living in the area. "My friends were all workers, but most were all steelworkers. We lived here for forty-four years. We all [other workers] bought homes owned by the Buckeye Land Company. Everybody in the area was tied to YS&T. People worked at YS&T because it was within walking distance."

While Jim DeChellis chronicled the neighborhood's development, Josephine ticked off a list of a dozen steelworking families who lived on or very near Wilson Street. These workers lived close together, and they had lived that way for as long as Josephine could remember.

The truth was that most workers stayed in Struthers until they died. The deep roots of area steelworkers were dramatically revealed by a study of YS&T workers who lost their jobs in 1977. The survey found that the "average steelworker had lived in the community for 29.9 years," and that "fifty-seven percent had been lifelong residents of the communities where they presently resided."[16]

Typical were those workers like John Skoloda who was born in Struthers on 11 January 1922 and was a "lifelong area resident." He worked for thirty years in the YS&T pipe department before retiring in 1971, and he died of cancer in 1993.[17] William Carol retired in 1985 after thirty-two years of service with Sharon Steel. When he passed away, William had been a resident of Struthers since 1945.[18] Frank Pozar came to Struthers in 1950 and retired as an inspector from YS&T in 1978. He died at age seventy-seven, having lived more than half his life in Struthers.[19]

Across the street from the DeChellises was the Zalusky family, who followed the DeChellises into the neighborhood. John Zalusky also worked for YS&T. Charley Petrunak and his wife, Florence, moved from a modest apartment into a home on Creed Street in 1962. Petrunak had a job on the railroad before finding a spot with YS&T, and his residence happened to be two houses down from where my family moved when we left Wilson Street. The Petrunaks raised three children on that spot and knew why the community was a strong one: "We all had a common problem. We all knew the other man's problems and needs. For example, if a guy on the street was working three to eleven, you'd tell your kids to shut up. If Mr. Blangero [a neighbor] was working eleven to seven, then don't go playing over there; play here. And his kids would come here. Mr. Peroga, the guy next door, your dad, Mr. Ritter, Worthingham were all steel mill people."

Republic Steel employee Armando Rucci moved to Struthers in 1958 and has lived on the same lot for thirty-five years. John Occhipinti was born in Youngstown and moved in 1952 to Struthers. He retired from Republic Steel in 1979 but still resides where he lived for most of his forty-three active working years, on Elm Street. Bob Dill came out of the service and went to work in the open hearth of YS&T in 1945. He moved to the Nebo section of Struthers and vividly remembered the closeness of steelworkers living in the community, "[We] drank together, partied together, went bowling all the time, visited each other's homes, and our kids knew their kids. We always talked about steel, no matter where we'd go. We'd go to ball games, take sandwiches, sponsored [union] golf and baseball teams."

The spaces workers occupied were carved out of larger, less intimate, and less meaningful places. At times the workers' community was independent of the wider society, and workers were often perceived less as national or municipal citizens than as residents of particular parts of town.

The southeast section of Struthers was known as Nebo. It was heavily populated by Croatian and Slovakian steelworkers. John Zumrick explained how this part of town got its name: "People in Struthers used to live on the north side. Well, the wind would blow the fumes, dirt, and everything from the northeast over the homes of the workers. So people moved across the river and built there, and when they realized the wind hardly blew that way, they called it 'Nebo' which meant *heaven* in Slovakian."

Two of the largest working-class, ethnic sections of Youngstown were known as "Lansingville" and "Brier Hill." Republic Steel's Youngstown plant was even referred to in the city paper as the "Lansingville Plant." Brier Hill was a predominantly Italian neighborhood famous for its "workingman's pizza," and more than one worker offered me his version of this local delicacy, baked dough with tomato sauce and parmesan cheese, cut into small squares. It was working-class food that could be bought at any pizza parlor in Brier Hill and found on the dining-room tables of Italian and Slovak workers throughout the area. Every Friday night my parents treated the family to twenty-four squares of pizza from Petrillo's Pizzeria. And if we had a coupon, we had thirty-two slices.

While neighborhoods often reflected an ethnic concentration, ethnicity was not the primary determinant of residence. Income level and housing costs tied to occupation were the most important determinants of where workers lived. The existence of ethnic and racially demarcated working-class neighborhoods was a result of industrial labor recruitment. Beginning with the first workers' homes constructed by YS&T between 1910 and 1930 until the company's settlement of Puerto Rican workers in the 1950s, residence had everything to do with class.[20] The primacy of class membership was in fact so important that it supports Cozen's claim that only "where similar

Charley and Florence Petrunak completely rebuilt their kitchen on Creed Street in Struthers. They lived two houses away from my family and across the street from a number of steelworkers' families. Standing in their own backyards, my mother and Charley often conversed about their respective gardens.

income levels and workplaces led immigrants to choose the same neighborhoods could ethnic concentrations emerge."[21]

Groups of workers all over the valley became anchored to distinctive territories. These spaces then assumed particular names and were individualized as separate, homogeneous neighborhoods of similarly situated people. Tony Modarelli, Anthony Delquadri, and my father identified with a neighborhood.

Modarelli said, "Everybody was in steel. I lived in Brier Hill where most people worked at the YS&T. It was a great neighborhood. People were very close and we all had similar lifestyles. Most of us went through the Depression together."

Delquadri added, "All my closest friends were from the east side and either worked for Republic Rubber or in the mill. My one and only home is the one we're sitting in. Right here in Brier Hill."

And my father replied, "We knew the neighborhood real good. Even in New Middletown there were steelworkers. Whitey Curtis and Frank Handle were there. Frank used to live in Lansingville."

The fact was that workers living in Struthers, Campbell, Loweville and Youngstown were likely to be family or friends, and the person living down the street or in the other room was probably a union brother. In the mid-1950s, it was very common to find two or more steelworkers living together. For example, three Agostino brothers worked for YS&T and lived together on Lincoln Avenue. YS&T employees Anthony and Pasquale Agnello lived within four houses of one another. George and John Adams of YS&T lived right next door, and three YS&T workers from the Massaro family lived on adjacent lots. The McClure and Dercloi families accounted for seven mill workers on the same small street.

On Wilhelm there were six Shevetzs and three Sedlackos working for YS&T. These two families alone made up 30 percent of the street's home owners.[22] Workers literally lived up against each other, most within three to five feet of another worker. John Pallay, an old timer from Republic Steel, emphasized that in order to paint the side of a house, a man had to prop his ladder against his neighbor's house.

A number of workers remembered that from their backyards they could watch a neighbor doing yard work two or three houses away. Given plots that were forty-five by ninety feet, working-class neighborhoods could be remarkably open spaces. Charley Petrunak learned exactly how open: "Mr. Damico once came down and fixed my lawn mower. He was watching from three houses up. Just came down through the backyards." My mother warmly recounted how on most spring days she would see from the back stoop, "Charley messing about in his garden." On more than one occasion they traded gardening secrets.

This intense aggregation of manual workers and machine opera-
tives in tight geographical areas ensured that community development would be a product of shared hands. Arthur Newell worked for
Sharon Steel, and his wife, Marcia's, description of life in the neighborhood is contrary to individualistic notions of private property:
"When our kids were growing up, we would get together in the
evening after the kids would go to bed. There were no fences in the
yards and you could move about easily but also be close enough to
hear if your kids were up."

Stable, interactive, and densely concentrated neighborhoods shaped
class practice within intimate locations. The meaning of community
interaction and the bonds of friendship were also often strengthened
by the interrelation of work-organized space and community institutions. In Struthers, for example, the high school auditorium was the
venue for premiering films about the local steel industry. Movies
produced by YS&T, focusing on the steelmaking process, were periodically presented to the public. In 1937, the city's Parent Teacher
Association sponsored the showing of "Iron from Mine to Finished
Product," and in 1961 a thirty-five-minute color film titled "Letter to
Youngstown" was presented.[23]

Community space in Youngstown was also used to rally worker
solidarity during the 1937 steel-organizing drive. South High School
was the location of a Congress for Industrial Organization (CIO) assembly, featuring speeches by United Auto Worker vice-president
Wyndham Mortimer and Steel Worker Organizing Committee re-
gional director Clinton Golden.[24] In addition, the workers of Local
1331 held social events at South High and also rented out the union
hall for weddings and graduations.[25]

## Ethnic Heritage

In some cases common national heritage drew workers together. A
local hall or club would serve as the venue for ethnic associations,
particularly for Italian, Croatian, Slovakian, and Puerto Rican workers. Ethnic meeting places like Krakusky Hall were among the best
known social clubs in Youngstown. Originally a Polish meeting
place, the hall was a social and political center for ethnic workers of
East European descent for more than fifty years. Class culture em-

braced ethnic traditions, including language, food, dance, symbols and belief. Worker affiliations with ethnic associations, however, should not be given undue importance. Only a few of the workers I interviewed regularly attended ethnic clubs during their working years. Many said that it was not until their retirement that they had the time to frequent such clubs. Some joined more out of reflex than of interest, and many were only occasionally active.

In 1937, at the age of ten, Armando Rucci immigrated to America from Italy. In addition to scores of unorganized family photos taken in America, he proudly pulled from a blue felt case photos sent from the funeral of a family member in Italy. As an adult he joined the Sons of Italy because his brother did, but after a while became disenchanted with the older members who "didn't want to change anything" and quit. Rucci was, however, serious about attending union meetings and made an effort to never miss one.

Certainly Youngstown Italian and Croatian steelworkers never considered that their nationality changed their working-class status. Most of the men I interviewed referred to themselves as working-class ethnics. Nearly every ethnic fraternity member was a steelworker. Rosters and attendance records of St. Anthony's Italian Club in Struthers, for instance, reveal that all but a handful of members were steelworkers.

While ethnicity did not shift class boundaries, race had the potential for stretching them beyond recognition. The character of Youngstown residential areas changed from the 1940s to the late 1950s. Housing settlements based on pre-middle-income wages produced more racially integrated neighborhoods than ethnically mixed ones. But after workers began to move into areas a bit further away from the mills, they found themselves living next door to families of different ethnicity. Rising incomes enabled workers of all nationalities to choose housing that broke down earlier ethnic barriers among neighborhoods.[26]

Bank lending policies left black workers behind in what were once integrated neighborhoods. The city of Youngstown was notorious for its residential redlining, and every black worker I interviewed expressed resentment about being unable to get a mortgage from local banks. Oscar Flemming, a Republic Steel veteran of thirty-seven years, told me that for some blacks it was necessary to travel to Pittsburgh in order to get a loan approved. "We couldn't move because banks wouldn't loan money to black folks to go into the suburbs."

The inequality in lending practices was evident in the furnishings of Oscar Flemmings's modest home. I wondered how a man who had washed up every day after work beside my father could have tolerated being treated so differently. Another black worker, Al Campbell, also recalled lending inequality. Campbell was the son of a Virginia coal miner and a member of the United Mine Workers Union. He explained how area capitalists restricted black workers from living in "white spaces":

> A white friend of mine named Chuck told me how he got a loan from the bank to afford a nice car. So I went down to the bank—Home Savings And Loan—because I made about as much as Chuck did. So the bank asked me about collateral. Did I have any stocks and bonds? At the time I was renting with my wife and she was also working. I said no, I don't have anything like that. He [the bank representative] couldn't help me. Well, later on I figured it was because I was black. You see, a black man couldn't get a loan in the city of Youngstown until the Civil Rights Movement. You could only buy a house through a land contractor, unless you were a GI.

While bank lending policies were the most egregious form of racism, Youngstown's downtown business class had a mixed record of welcoming blacks.[27] Oscar Flemming recounted his own experiences with bigoted merchants:

> [Downtown] store owners wouldn't let blacks use the bathrooms. We were watching the parade and my daughter, who was a baby, needed the bathroom but she wasn't allowed in Strouss' or McKelvey's, but little white girls were going in and out. We had to carry a potty chair around with us. Some restaurants and theaters were segregated, too. And they didn't want blacks to swim in south-side pool. You had to come over to Lincoln Park. The city also had segregated public swimming pools, which became the sites for a couple of notorious racial encounters.[28]

In Struthers, blacks made up only 2 percent of the population. Ninety-five percent of these people lived on the north side, while nearly all the Hispanic steelworkers lived on the east side of

Youngstown and in mixed ethnic sections of Campbell.[29] Black residences in Struthers were located primarily between Yellow Creek
and the north side of the Mahoning River. Census tract data reveal
that nearly 90 percent of the town's black population lived in tracts
105 and 108.[30] This explains why my only contact with the sons and
daughters of black steelworkers occurred during my high school
years.

I cannot recall an instance of prejudice in my family, but I starkly
remember an act of which I am ashamed. Although race had nothing
to do with the incident, I soon realized how race could complicate
class identity. In the heat of a tense intramural school basketball
game, I knocked down Artie Bledsoe, a black player, in order to
thwart a layup. He missed the shot and badly injured his knee. He
had to be carried off the court and immediately prepped for surgery.
For days at school, rumors spread about a race fight that would occur
to get even for my overzealousness. Perhaps as much out of fear as
out of guilt, I paid Artie a visit at his home on the north side to apologize, and, as a result, there was no race fight. What I recall about the
incident now is the surprise on Artie's father's face when he found
me at the front door. That look of surprise was not there because I
had come to apologize but because someone like me had come at all.

There was considerable residential segregation, and Struthers's
black population in areas other than that between Yellow Creek and
the north side of the Mahoning River was sparse. But 50 to 75 percent of the people living in tracts 105 and 108 worked directly in
steelmaking, and it seems unlikely that workers of any nationality or
race would not interact. Even racially oriented housing patterns did
not automatically preclude social interaction among workers and
their families.

Segregation, where it existed, was not absolute. While the postwar
community of workers outside the steel mills did not equally include
white, black and Hispanic residences, all of the white workers I interviewed noted that they had black high school classmates and
black friends in the neighborhood. This is consistent with what
black workers recalled about the years before the advent of Veteran
Housing Administration loans. By the mid 1950s, most racially integrated areas had become noticeably segregated.

In 1960, 45 percent of Youngstown's census tracts had at least a 20
percent nonwhite population. While blacks were completely locked
out of the west side of town (only 1 percent), they were almost

evenly dispersed across the city's remaining areas.[31] Consequently a
person could often be distinguished more by location than by race.
John Barbero, a YS&T worker, said, "It surprised me during World
War II that whenever I ran into somebody from Campbell he had this
psychology. He was always fair in a question of discrimination,
whether to blacks or whites, because the whole town was this way.
Even to this day, when I run into that generation of people out of the
city of Campbell I automatically know how they are going to feel."[32]
While segregated housing patterns that evolved after World War II
did substantially reduce community alliances, race did not necessarily hinder a worker's identification with class.

## Communities of Steel

Class identity was in part constructed out of being from a place
where steel was made. It is remarkable to note that in an area with
approximately sixteen thousand steelworkers, for much of the 1960s,
fewer than 1 percent lived someplace other than in Struthers, Campbell, Loweville, and sections of Youngstown.[33] In contrast, company
executives made their homes in the richer, more modern suburban
towns that grew up further from the Mahoning River. Turn foremen
and supervisors accounted for less then 2 percent of Struthers's mill
employees.[34] In the communities where workers lived, there was
limited opportunity for worker and boss to socialize.

Whole communities made steel. In Struthers, for example, the
community's reliance on steel production was creatively dramatized
in a citywide celebration held in the summer of 1948. To affirm
Struthers as the "cradle of steel west of the Alleghenies," the city
turned out for seven consecutive days to witness the "Cradle of Steel
Pageant."[35] The proceedings included a beauty queen, an old-timers
baseball game, bowling tournaments, fireworks, a watermelon contest, parades, a soapbox derby, street dancing, and a "Bombing of the
Valley" by a squadron of private planes. The week's festivities were
capped with a pageant performed at the high school football stadium.
An anonymous poem printed in the pageant program conveyed the
unity of this workers' community:

> Oh give me a job where the smelters roar,
> Where the fiery monsters gulp the ore

> Where the thundering furnaces rock the earth
> And labor hard to bring the birth
> Of steel, America's natural cream
> That sputters and flows in a white-hot stream;
> There where the base of our wealth is laid,
> There let me work at my job and my trade.[36]

Nearly anyone in the community could have expressed the same sentiments. Steelworkers were united through the way shop floor relations and community associations reciprocated. "Closeness started in the mill and then got carried out," U.S. Steel employee and life-long Struthers resident John Varga told me. "Guys depended on each other at work, this carried on out of the plant so that you could trust these guys."

## 2  Santa Claus Was a Steelworker

Every Monday night for many years my mother bowled with a group of women. Like her, a number of them were married to steelworkers. Mom was a pretty good bowler and proudly displayed the trophies she had won in various leagues. Her bowling prowess even became a friendly source of kidding with Dad. My father also took to the lanes at the Holiday Bowl. The Bowl was a convenient place for him to compete because it was located at the top of our street. My parents preferred team-league bowling to individual competition. Team bowling was a very popular form of social activity in the Youngstown area because it was cheap and involved friends and workmates. Dad and Mom obviously enjoyed bowling, but what was more important, they said, "We just wanted a chance to do something with people we liked." One eight-member team included Dad, Bill Mastrangelo, Kenny Wanamaker, and Lou Veltre. All of them were steelworkers at one time or another.

In the steel towns of the Mahoning Valley, class ways of life took shape in a variety of diverse community relationships where workers were recognized as honorable and virtuous men. Most steelworkers had lived in the Youngstown area all their lives, and they were usually continuing an occupational lineage that went back two or more generations. They were known in town as "good men," "hard working men," and "solid family men."[1] The dense networks among workers are the subject of this chapter. It specifically examines three recurring forms of social interactions: (1) unmediated, direct residential visiting, (2) organized clubs, teams and churches, and (3) unions. I treat these forms separately, although they overlap considerably as they take place. While the social connections among workers were influenced by ethnic and neighborhood loyalties, for the most part they did not include non-working-class people.

## Neighbors

During the course of daily interactions, relationships among the steelworkers usually evolved through informal face-to-face contacts. For Jim Visingardi socializing was commonplace and casual: "[We] got together at union picnics and sometimes at different guys' houses, maybe five or six families at a time. In fact, we had a card group that rotated around different homes." Social life was no different for my parents, who found that material conditions often determined the context within which friendships developed. When asked about common associations, they explained, "We played cards with Kenny every weekend. That was our recreation. That was a cheap night. We would take turns at different couples' homes, make sandwiches, have chips, drink coffee. It was great, just great."

Card playing was a popular way for workers to mingle. Mary Ross was one of the many women who entered the mills during the war. She had many friends who did likewise, and they all worked together in the "spike plant." Ross had been married twice—both times to a steelworker—and out lived both husbands. During our interview, we sat across a table decorated with candles and surrounded with photographs, and paintings of Croatian women in native dress. Before entering the plant Ross and her friends started a neighborhood custom. "I was a member of a card group before the war," she explained. "We played once a month at different houses and there was eight of us." The war interrupted their games, but after it ended they "got back to playing and the group stayed together until the '70s."

Whether they played cards or just visited, workers found comfort in the presence of one another. It was no different with Armando Rucci. He and his wife "would visit all the time. [They] usually just sat and had coffee and chatted." Tony Pellota made the point that what workers did together they did in the most spontaneous ways.

Armando Rucci, like other workers, was fond of meeting with family and friends around the dinner table.

"We did visit each others' homes and we'd go out to pubs and restaurants." But according to Pellota it was "nothing big or formal."

Joe Ryzner lived a little differently than his co-workers. For many years he owned a home in rural Liberty Township, far from the working-class suburbs of Youngstown. There were a few steelworkers around, but most of the neighbors were farm workers. Ryzner loved his old neighborhood because of what he and others did together during special times of the year: "We had a lot of fun, picnics and fireworks all the time. The kids were everywhere and you were everybody's parents. We would Christmas carol together. In the fall, we would pick walnuts and hickory nuts and butternuts. One year we loaded that truck with so many nuts that the springs were flat. There were a lot of good working people living there. People were sociable. Nobody had much and we were all pretty equal."

The social worlds of work and community often merged.[2] The technical requirements of a particular job usually introduced workers to one another. What began as cooperation on a task would often expand into departmental solidarity. Workers in the electric weld department at Republic Steel, for example, were recognized throughout the plant as being a tight-knit group. Workers who never worked "the weld" told me of how "family-like those guys were." Chris Cullen worked for Republic Steel, but never in the weld department. Nonetheless, he said, "The electric weld was the closest [department]. I heard when I started working for Republic that if something happened to one guy—that's it, they'd all walk out."

Oscar Flemming put in some time in the electric weld department, and when he got there he was surprised to find, as he said, "I knew every guy working there. All these guys were from Wabash, Franklin, Cyprus, Gibson [Youngstown city streets], all from the neighborhood. They ran the electric weld." When separately asked about their closest friends, electric weld workers all mentioned one another. Better than anyone else I spoke to, Armando Rucci exemplified the special bond he formed with other members of that department. In his remembrances he stressed, "If you work your lifetime with a bunch of guys they become your family. In fact, truthfully, I spent more time with the guys in the mill than with my own family. I'll tell you, they become part of you. If one of my buddies cut his finger I felt the pain."

My father was known to Rucci as Bob, but when Rucci wanted to communicate how close he felt to the men he worked with, he referred to my father in sweet terms: "your daddy." Becoming "part of you" was a process workers often began on the job and then extended into wider social circles.

Workers pointed out that a good neighbor or community member was someone who recognized when the guy next door needed a hand. Chris Cullen from Local 1331 expressed this in the understated, unassuming way that most workers did: "You did whatever you could. You spent a great deal of time helping people with projects around the house. People pitched in; just give a guy a call and he'd come down." In most cases all that was needed was a request for help. Bob Dill always knew, "If you needed a truck someone would come over, and some guys had gardens and they'd bring you vegetables. If you needed dirt or some carpentry, people would come over and give you what they had."

Coming over was easy enough when you lived, like Brad Ramsbottom, surrounded by other steelworkers. He and I talked for two and a half hours on his front porch. He confirmed the cooperative nature of workers' community relationships by explaining, "A guy would say he needed a patio or driveway put in, and we always helped. We did that all the time. I spent a lot of time helping guys move. We especially helped one another with their kids." As Ramsbottom and I moved from one subject to another, it was clear that he was proud of the community.

Charlie Harp used few words to say what all workers accepted as a given: "Guys helped other guys build their homes. Anything that came up to help one another, you were there." Sometimes the help was the difference between life and death. Anthony Delisio's wife, Joanne, offered a poignant recollection: "My husband was a good man, but he had a drinking problem and got into a lot of car accidents. The guys would be with me all the time. They came to the house, went to the hospital, would pick things up for me. Once when Anthony had a bad eye infection they brought me a check to buy things for the home. And these weren't only the guys my husband grew up with; people from over the mill were very sensitive and helpful to me."

Help from another worker could also mean more than giving a few bucks and a helping hand. Sam Shapiro had coronary bypass surgery, and "five guys from the mill gave blood." He recalled, "Some of the guys came to my wife at home to help, but a cousin who lives up the street wouldn't do a thing to help me and my family." Shapiro then echoed the statements of many other workers: "My best friends were from the mill and when you needed help they were there. They were goodhearted men who knew each other and trusted each other."

Sometimes that help was directed toward someone other than the worker. Lillian Bergman will never forget the help she got from the neighborhood women:

> I was pregnant with my first child when I began to feel pains, so I called Dolores [the wife of Joe Vlosich]. She said, "Oh my God, you are so many minutes away from labor!" Tom [her husband] had been laid off for weeks but was called the previous night to come out to work at 6 A.M. So the neighbors came to the door with Dolores and her sister Barbara and Barbara's daughter, who was studying to be a nurse. They put me into the car and drove me to the hospital with a police escort. We got there on time. Tommy was born August 16, 1960. Now after the baby I couldn't come home, so the neighbors came over and made some meals and helped my husband.

Relations among these working-class families were deeply embedded in community interaction. Steelworkers' families not only knew the people around them, but knew their needs.

## Faith and Heritage

While the most intimate working-class relations were predominantly direct and unmediated by organizations, there were important social associations that corroborated the workers' sense of class. Workers joined organizations that were established primarily to serve the residential community, but because of their membership they became specifically oriented toward a wage-laboring class. In the city of Struthers, St. Nicholas Roman Catholic Church exemplified the working-class focus of community organizations.

St. Nicholas Church had so many steelworkers in its congregation that the parish held a special Sunday morning sunrise service to accommodate workers who had to be at work at 7:00 A.M.[3] Jim DeChellis was one of those workers who made it to Mass despite rarely having the Sabbath off. DeChellis noted that "most steelworkers were good churchgoers, and St. Nicholas had a 6:00 A.M. Mass on Sundays for guys who were going to work that morning." Workers would come in their street clothes and in their work blues. Dress didn't seem to matter much, and "the Mass was always very well attended." DeChellis took special pride in the fact that "the church was very sensitive to the mill workers."[4]

The church was a working-class house of worship. Holy sacraments were administered primarily to steelworkers, and their families were often blessed with the sponsorship of fellow class members. Evidence of the congregation's working-class composition and of community participation in receiving the sacraments can be found in the baptismal record. In the years 1948, 1950, and 1952 there were 490 baptisms performed and 53 percent of them involved a steelworker either as father or godfather. For the year 1950, 8 percent of the baptisms involved steelworkers baptizing the children of other steelworkers.[5]

Working-class relations within community churches were not, however, limited to spiritual offerings. The priests at St. Nicholas were fond of walking through the neighborhoods and visiting working-class parishioners. My father remembered, "Father Gubser

Jim DeChellis, like many other Struthers steelworkers, was a faithful member of St. Nicholas Church. It was very common for steelworkers to serve as best men at weddings for other steelworkers and as godfathers and confirmation sponsors for the children of other steelworkers.

would come over and read his book [Bible], usually when we had company." The parish rector, Father Petric, was fond of interrupting street football games to toss a pass or two. My youngest brother, Rick, was often the intended recipient of Father Petric's efforts, but apparently he dropped more throws than he caught. When friends and families packed the church to witness a confirmation ceremony, Father Petric referred to Rick as "butterfingers." When Rick was called up to receive his confirmation papers, he bobbled the parchment, but hung on to complete the sacrament.

Church activities also brought workers together for dinner, dances, fellowship, and pastoral purposes. In Campbell, St. Rosa de Lima was the home church for thousands of Puerto Rican steelworkers. St. Nicholas Church in Struthers and Mount Carmel Church in Youngstown sponsored festivals and dances for their many parishioners.[6] Mount Carmel was an old Italian church that attracted Italians from all over the Youngstown area. My uncle Frank Frattaroli belonged to Mt. Carmel's Vestibule Club and Holy Name Society. His wife, Annie, was vice president of the church's Mother Crucifix Society.

Although many Struthers steelworkers appeared to be confident supporters of St. Nicholas Church, it does not appear that the pulpit was often used for class mobilization. Workers did not recall the priests ever talking about the mills or directing sermons toward class-related issues. They claimed that the priests were always neutral during strikes. But the church did help workers overcome the material strains of long strikes and periods of class resistance by announcing "that if you had a voucher from the union you could go to St. Vincent DePaul for food."[7] A chapter of the St. Vincent DePaul Society was organized in April 1959, and "second collections" were taken at church services to aid the relief fund.[8]

When the valley suffered the devastating blow of steel mill closures, local churches held significant influence with Youngstown area steelworkers. St. Nicholas Church urged its congregation to "remain calm" and to "pray as a community" because the "power of prayer is very much needed." The church encouraged Sunday Mass attendees to attend a community meeting about the plant closings and reminded everyone, "We need each other—we need to be our brothers' keeper."[9] The church also held a community Mass on behalf of the Youngstown Sheet and Tube (YS&T) families, celebrated by Youngstown Dioceses auxiliary bishop, William Hughes.[10]

While the pulpit was not used for direct mobilization of workers, the most comprehensive and sustained political response to the area's mill shutdowns of 1977–79 came from the local clergy. The formation of the Youngstown Ecumenical Coalition, a tripartite organization of religious leaders, constituted the workers' best hopes for economic renewal. The coalition's efforts were multidimensional, and they began with a pastoral letter designed to "renew the ties of common purpose and concern which can help us to become a better and more just community."[11] While the post-shutdown period is not the immediate context for this book, the language used in the coalition's letter does reflect the nature of the relationship between an interfaith community and its working-class members:

> This decision [mill closing] raises profound issues of corporate responsibility and justice. . . . We say this because the decision is the result of a way of doing business in this country that too often fails to take into account the human dimensions of economic action. . . . Some maintain that this decision is a private, purely economic judgment which is the exclusive prerogative of the Lykes Corporation [parent of YS&T]. We disagree. This decision is a matter of public concern since it profoundly affects the lives of so many people. . . .[12]

Workers were aware of the coalition's purpose and appear to have taken this letter seriously.[13]

While the juncture of faith and secular practice influenced interpersonal relations, many workers belonged to social as well as religious organizations. Fraternal associations, which varied according to nationality and purpose, were made up primarily of working-class people.[14] Republic Steel worker George Porrazzo stressed that "the VFW, the Croatian Club, the St. Lucy Club, the ITAM (Italian American Club) were very popular. Guys were members of a lot of the same associations." Jim Visingardi was a member of one of the oldest associations in Struthers St. Anthony's Society. He was proud to point out that he was a member since 1948, and that "maybe 99 percent of the members were steelworkers and nearly all Italians."

There were a large number of Italian workers living in the communities built in the low-lying areas near the Mahoning River. In 1960, the population of the city of Loweville was roughly 45 percent foreign-born Italians and their children. In Struthers, Italians represented 35 percent of the population.[15] Of course, Italian settlement

Jim Visingardi's father was a YS&T employee and a darn good accordion player. He often turned the cramped quarters of the family basement into a music hall.

in both areas was closely linked with steelmaking. The first Italian couple in Struthers was Felix and Floradena Carbon. Felix Carbon was employed for forty-six years with YS&T before his retirement. Agostine Vesparaian, a mill hand and the Carbons' neighbor across the river, was the head of Loweville's pioneer Italian family. He immigrated to the United States in 1882, and, like most Italians who came to Loweville, he arrived at the Pennsylvania Railroad Depot. Vesparaian worked for forty-five years as an employee of Sharon Steel Hoop (later Sharon Steel). He married Antonin Carbon, and their sons Carmen and Frank both labored in area mills.[16]

The large concentration of Italian workers supported local chapters of the Sons of Italy. Anthony Delisio (Sharon Steel), Lee Delsignore (United States Steel), Mike Scocca (Republic Steel) and Anthony "Red" Delquadri (Republic Steel) were all members of different chapters.

During the peak years of Italian immigration (1902–1924) approximately one hundred families settled in Loweville. Immigration in the valley was so intense that "numerous Italian communes [were established] within a radius of fifty miles to the East, West and South of Loweville."[17] Family expansion generated community networks that intertwined one kin group with another. It was common practice for a local Italian resident to be a bridesmaid, best man, or godparent to another community member. By 1975, the reciprocal exchange of sponsorship resulted in 95 percent of the town's Italian families supplying either *commara* (godmother) or *compare* (godfather) to another family.[18] Armando Rucci of Struthers, in fact, "baptized a couple of steelworkers' children."

Of course, the practice of steelworkers baptizing other steelworkers' children was not exclusively Italian. YS&T employee John Mikulas was godfather to Joe Flora's son, and Ramon Ramirez's six children were all baptized by "friends in the mill." Ramirez was deeply influenced by the Catholic faith, and his religious observance only strengthened his working-class relations. He volunteered at St. Rosa de Lima Church in Campbell during the annual spring festivals. The church's congregation was primarily Puerto Rican, and most of them were steelworkers. Ramirez pointed out, "I was asked by two couples—the guys worked in the mill—to baptize their children, and my six kids have godfathers from the mill."

As we sat on his front porch sipping strong Puerto Rican coffee in the August heat, Ramirez recited the names of each of these godparents. He was eager to talk about his "mill pals," but after a brief story about one favorite godfather, he insisted that we stop for a moment and see his garden. In the back of his house, crafted on a modest slope, was a diverse garden of potatoes, peppers, onions, and eggplants. When I left Ramirez's home later that day, there were two thing he had taught me: work, church, and friends make a harmonious life, and a good eggplant needs a moist soil.

Women's associations were common among local religious and ethnic groups, and what most had in common was the degree of leadership displayed by the wives of steelworkers. The Struthers chapter of the Sons of Italy had a very active auxiliary organized predominantly by the wives of steelworkers. The nine-person executive board of the Italian Mothers Club was composed of seven steelworkers' wives. They met once a month and planned social outings for both men and women. One of their more popular activities was an annual sausage fry regularly attended by a large number of community residents.[19] The Dames of Malta was run by a directorate of eight women, seven of them married to steelworkers.[20] The social committee for the thirtieth anniversary of the Wide Awake Temple consisted of sixteen members of the Pythian Sisters; ten were the wives of steelworkers.[21]

Steelworkers' wives also dominated the auxiliaries of various local VFW Posts. The auxiliary for the Italian-American World War Veterans, Post 5, was run by women from steelworking families.[22] In addition, the Ladies Auxiliary of the Fraternal Organization of European Veterans, Post 2536, was heavily represented by steelworkers' wives.[23] Working-class families were also well represented on educational associations. The Sexton Street PTA recruited women to be room mothers, and out of nineteen appointments, six of them were the wives of steelworkers.[24]

Ethnic associations were also prevalent among Puerto Rican workers living in Campbell and Youngstown. Between 1950 and 1952, a significant number of Puerto Rican men went to work in area mills.[25] The majority found employment with YS&T, and most Hispanic workers bought mill homes on the east side of Youngstown and in the city of Campbell. While Puerto Rican citizens made up only 5 percent of Campbell's population, 80 percent of them lived in one small triangle bordered by Penhale, Wilson, and Thirteenth Street.[26] Immediately to the north of this area is the east side of Youngstown, home to 46 percent of the city's Puerto Rican population. Notably, 67 percent of the county's total Puerto Rican population resided in only six adjacent census tracts.[27] Within this close compound of square, brick-framed mill homes, many Puerto Rican workers quickly formed chapters of social service organizations. Ramon Ramirez, Cayetano Caban, Augustine Rodriguez, and Porfirio Esparro were all YS&T workers, as well as members of the Sons of Boringuan.

## Leisure

Most Youngstown steelworkers were avid sports fans and often associated with one another at local athletic events. Local high school football, basketball, and baseball games offered workers a chance to engage one another in a favorite family and class subject. Sons and daughters who played sports were a common topic of conversation among workers. Participation in sports encouraged workers to establish a deeper group identity through their weekly association with

athletic children. In fact, the initial bridge into a work- or
community-based friendship was often through local athletics.[28]
John McGarry knew the families of other steelworkers, and wherever
the community's children played, workers reconstituted parts of
their workplace relations. McGarry said,

> The majority of your life you interacted with each others' families.
> You would go to football games and baseball. The girls were cheerlead-
> ers and there was little league. I also was active in the Campbell ACs
> [Athletic Clubs] and supported the Class B baseball team. A lot of steel-
> workers were in both organizations and many had sons who played
> baseball for Campbell AC. We had baseball picnics with all the fami-
> lies every summer. You went a lot of places with your kids where you
> saw other steelworkers and your kids usually had friends whose father
> was a steelworker.

Republic Steel employee Tony Perry was a member of the Hubbard
Boosters and remembered, "workers were big basketball and football
fans. Most of us had kids who played ball and we'd spend most Friday
nights in the fall watching football. A lot of us were members of the
ACs. Sports would bring you together when you were out of the
mill." Brad Ramsbottom was not only a member of the ACs, but also
active in the Struthers Gridiron Club.

Knowing a worker well meant knowing whether his kid played
sports. The son of a man that Tony Modarelli worked with "had been
in the hospital for a couple of days with leukemia. He was given a
physical before a football game for Youngstown State. Vince Pasco-
vitch was his name; the kid had been a good quarterback for Fitch
High School."

Athletic contests were opportunities for workers to relax, get away
from work, and engage with family and friends. They were also an affordable, working-class activity. Most steelworkers worked rotating
shifts, and this fractured the ways in which they participated in the
community. Constantly irregular schedules made it difficult to plan
family trips and attend children's functions. Tom Bergman's wife,
Lillian, explained "We could never take vacations with the kids. We
couldn't afford to go away, and with Tom working different turns it
was impossible to ever plan things with the kids."

Rotating work schedules meant that often a steelworker's wife
would be the lone parental representative at a child's game. Little

league games attracted a predominantly female crowd, and a high school football crowd may have excluded a son's father, but rarely his mother. While I remember my father attending many of my baseball games, it was my mother's presence near the outfield fence that I could always count on. Mom was an obvious figure at those games because she had a distinctive head of blazing red hair. She was a very sophisticated fan who, being the mother of four boys, was torn between wanting to keep her sons from being hurt and feeling downright insulted if they were removed from the contest.

Lillian Bergman, however, never quite understood the rules of either baseball or spectatorship: "The other women at the game would get mad at me for cheering. This I didn't understand. I thought you were supposed to cheer when a player does something good. But they had to tell me to only cheer when our boys do something good. And another thing confused me. I heard a woman say that the boy had 'stolen a base.' I was very upset and said, 'Then he must give it back!' I didn't know that stealing a base was a good thing."

Most of Tom Bergman's time off was spent watching his three sons play baseball. As we sat through a much needed summer thunderstorm, Bergman reminded me that more than ten years earlier I had coached two of his sons. There was little about those games that he had not committed to memory, and he beamed about what was a special vacation: "I spent the most enjoyable thirteen weeks vacation watching the kids play baseball. That was the best thirteen weeks of my life. Everyday I took the kids to the ball field and watched both Mike and Tommy play in different places. I really enjoyed that. It was a chance to be with the kids, and it was wonderful."

The mingling of steelworkers' kids through team sports was as extensive as John McGarry suggested earlier. In 1956, sixteen players were picked to the Struthers Little Boys League (SLBL) All-Star Baseball Team. More than half of the boys picked were the sons of steelworkers. Between 1956 and 1959, steelworkers' sons made up 40 percent, on average, of SLBL teams (ages 8 to 12). Furthermore, the proportion of working-class participants in organized sports increased with age. In the town's Teen-Erly Baseball League (ages 12 to 15), 56 percent of the participants appear to be from steelworking families.[29] The 1959 high school squad, the Wildcats, had nineteen players, and twelve of them had steelworking parents.[30]

Steelworkers' children were often coached by other steelworkers. Republic Steel's Joe Opsitnick coached Thatcher Heating in the

SLBL, and YS&T employee Tony Karis helped out with the Nebo Indians.[31] Steve Elash also spent one summer coaching the Teen-Erly League All-Stars.[32] To do this, he had to forgo any family travel. YS&T worker John Lilak went one better then most workers: he served a few years as president of the Teen-Erly League.[33] George Bodnar was a little league coach in Struthers, and he could still remember the names of at least half a dozen of his players whose fathers worked in the mills. In the summer, he was usually very busy with baseball practices, games, and tournaments. George also had an occasional second job, but somehow managed to fit in baseball. "On days I didn't do the second job I would run home [from the mill] throw water in my face, grab a cup of coffee, and run to the ball field."

Most workers also socialized in less competitive forms. Workers easily found space to hold dances and parties. The Elms Ballroom was a popular place for weekend evenings.[34] Mary Ross spent a good deal of time with her friends at "Croatian socials, dancing to the tamburitzas." In the late 1960s, Boyd Ware and Charley Harp spent so much time with other workers that they decided to form their own African-American social organization, the Six Rocket's Club. "We would rent a hall and sell tickets. We rented the union hall for dances and parties. We'd get two hundred people to come. We'd sell the tickets at work and in the neighborhood. You brought your own food and bottle."

Engaged in clubs, card groups, teams, taverns, and seasonal events, workers encountered each other in seemingly incongruous places. Ramon Ramirez was equally likely to see a worker "at the church and the pool hall," and occasionally during the course of these activities workers came into contact with people from other classes. Red Delquadri made an unflattering distinction between the different economic classes represented in an organization to which he belonged: "I joined Ameritol [American-Italian]. It was a civic club, and we held parties for children who needed shoes or were handicapped." I questioned him, "Did the club include people from different economic classes?" He replied, "Yeah, but the big business guys were there for business reasons and for political gain."

## Union as Common Ground

The workers I interviewed gave the impression that life's most important moments were almost always shared with other steelworkers. In essence "workers learned everything, including what they believed, from working, talking, and living with other workers."[35] Tony Pellota worked most of his adult life at Sharon Steel and earned the degree of success that accompanies longevity. It was essential to note, he said, "Every time I made a move it would lead to a better job. You know, it was a joy to go to work. In the old days, everybody was like a family."

Cayetano Caban claimed that he "learned the most in the mill," and because of what he did with other workers, he said, "I could give my children an education." John Zumrick's most influential experiences consisted of voluntarily sharing his "bonus with guys who needed extra money." "Nothing was more important than being a worker—a worker over being a Catholic or being a citizen or being a Croatian."

Workers consistently spoke to their common condition and the way it generated familial affection. Charley Petrunak emphasized that working-class attachments were formed through hardship. "That's right, hardship. Hard times brought you close. It made you a good person. You couldn't do anything by yourself. You had to pull together. There was nothing you could do individually." Workers were unanimous in their dissent with an autonomous, rugged individualist path to success. When I asked whether their personal efforts, resourcefulness, and commitments would have been sufficient to lead a quality life, they responded as Joe Ryzner did: "No way. You needed other workers to move ahead. It was the only way to go."

While racial and personal differences were never eliminated, workers labored "in the same conditions," "had to work for a living," "were in the same boat," "had a similar kind of way of life," and "worked at the same level."[36] John Zumrick summarized it simply by stating that workers felt "love for one another." Sensing that the word love seemed a bit melodramatic, he explained, "you got along because you worked together and that made him [another worker] a brother to me."

In cases where workers were not individually capable of overcoming ethnic, cultural, or geographic inhibitions to interacting with one another, the mechanisms most likely to unite workers as a large group were the plant itself and the local union. The absence of overt racial attitudes did not automatically lead to integrated friendships. Some white and black workers did socialize outside the plant, but a more common situation was described by black bricklayer Al Campbell, who spoke positively about many white workers. The majority were not, in his eyes, "race haters." But he said, "I didn't see my white friends when I left the plant. . . . One day a [white] mason came up to me and says, 'Albert, you have to start spending time with the other bricklayers.' I said, 'Those guys over there [black bricklayers] are my friends. When I leave here and go across that bridge I don't see you guys.'"

My father concurred with Campbell's description of residential divisions based on race, but stressed that the social geography at work could make separation difficult. He pointed out, "Black and white workers had lunch together and talked about sports and family. . . . you have to understand that inside the plant there was no place to go to be alone or be separated."

Inside the plant workers were workers in spite of racial identity. The job required coordination, and workers were quick to appreciate the need to cooperate. If workers were going to prosper a degree of solidarity was essential. Despite racial bigotry, black and white workers were unanimous in believing that the steel companies saw them as only "check numbers." Among the black workers that I interviewed, Charlie Harp and Boyd Ware were probably the most angry about white racism in the mill.

Ware was a thin man with sharply chiseled facial features. He spoke slowly and pointedly about racism. He was hard pressed to find anything in common with his white co-workers. Yet when it came to interpreting how the company perceived all workers, Ware agreed "The company didn't care what color you were long as you made them some money. They still treated you like workers!" Black, white, and Puerto Rican workers pointed out racial differences, but when I asked what the most important thing was that they had in common with other workers, the near unanimous response was "We work for a living." Ware and Harp also agreed that all workers, black or white, were check numbers to the company. Not surprisingly, the phrase "check numbers" or something similar was also used by most white workers to describe what the "workers were to the company."

Willie Floyd agreed that discrimination came from both the company and the union. But he also stressed that, "if the union was flawed by racism, it was at least against the company." Even a color-bound union was a better bet than a profit-driven company. Hence, united as wage laborers in the production process, "working for a living" meant joining the union. Membership bought a chance to make good. Curiously, Floyd could recall the names of good white union members, but he failed to reproduce a single foreman's name. When Floyd retired, YS&T gave him a small, glass-encased clock. At the time of our interview, the clock had been broken for some time. He saw no need to fix it.

Charlie Freeman was a black laborer who rarely interacted with white workers outside of the plant, but he indicated that integration did occur at "union picnics; everybody came out to those." The union could function as a bridge to the community of class. Mario Crivelli admitted that not every steelworker independently associated with other mill hands. "Unless the union brought us together I didn't spend much time with other guys."

Except for strike action, the most common reasons for a union-called gathering of workers were to celebrate the holidays and to enjoy a brief respite from the mill. Workers proudly remembered that union-sponsored picnics and Christmas parties were well attended. Workers spoke very enthusiastically of the all-day picnics held at Idora Amusement Park. Each worker would be given a "United Steel-workers of America, AFL-CIO Free Ride Pass" for the entire family. For Christmas, Union Locals would announce a "Registration for the gala Christmas Party" and urge all workers to "register [their] children to be eligible to receive Christmas Gifts."[37] George Porrazzo placed the union hall at the center of these collective affairs. He recalled, "The union had outings at Idora Park and Christmas Parties. Guys would bring their families to the union hall, and someone would dress as Santa Claus and give out gifts."

Local 2162 in Struthers held an annual Christmas Party at its I. W. Abel Hall. The local provided food and gifts to over 550 children.[38] It was also a common practice for every local to hold a holiday "grab bag" for its members. In good years such a practice was a nice gesture with little significance beyond the additional ability to provide a basket of gifts for a child. But during the all too frequent lean years, these gifts taken home from the union hall often represented the only ones given on Christmas morning. Oscar Flemming counted on this: "The union had Christmas and other parties where you could get toys for your kids." If not for the union "grab bag," or the assistance of union friends, some years Santa Claus would have skipped over the houses of Youngstown's steelworking families.

I expected to miss Santa at least once. In 1959 the steelworkers waged a long drawn-out strike. For months, my parents had not seen more than a few cents in a supplemental income check. The check, needed as it was, rarely got cashed. With much incredulity, my mother pointed out that the weekly "check for sixty cents wasn't worth enough for the thirty-five cents it cost to drive downtown and park the car in order to sign for it." We lived, as most other workers did, on the credit of considerate middle-class merchants. But asking for assistance to purchase such necessities as clothes and cuts of meat was one thing, while Christmas toys, even for a four year old, were a luxury. My parents explained that Santa would probably not stop at our house. It wasn't anything I had done. Next year he would leave twice as many gifts.

Santa needn't have bothered. Christmas morning I raced to the tree standing in the living room. Underneath was a box wrapped with shiny paper. My name and the words "From Santa" were written on the top. Within seconds I had the box open. What I pulled from it was a red truck with a front-loader bucket. The gift had come from the union. Dad had participated in the grab bag. Years later I found out that nearly every Christmas he had come home with a union donation for each of his sons. Dad recalled a "cowboy gun and holster set" as being a particularly nice gift. But for me, no present ever matched the gift I got when I only got one.

Local 1330 went a bit further to enhance Christmas for union families. Beginning in 1951, the local sponsored two Christmas shows at Schenley Theater. The performances were open to kids from two to twelve years of age and attracted large working-class audiences.[39] The local also held a more conventional party at the union hall, and for one fifty-cent ticket a worker could bring all his kids to the affair.[40]

During the summer, the Mahoning County CIO Council sponsored all-day discount outings at Idora Amusement Park.[41] Workers and their families would come from all over the county to spend a long day playing games of chance, riding the Ferris wheel, and eating cotton candy. While annual field days were well attended, no occasion drew a bigger or more energetic show of workers than Labor Day. The first citywide Labor Day celebrations for steelworkers were sponsored by the YS&T Company. These events were undoubtedly designed to foster goodwill and forestall a CIO-Steel Workers Organizing Committee union drive. They revealed a company that believed it had the unquestionable loyalty of Youngstown residents. In what can only be described as a monumental show of arrogance, the company organized a Labor Day celebration less than one month

Every Christmas my dad brought home gifts from the union hall for me and my brothers. Dad especially liked the gun and holster set.

after the disastrous end of the 1937 Little Steel Strike. The company expected thirty thousand people to attend.[42]

Workers obviously found the steel companies' generosity less than sincere, and within a few years after the United Steelworkers of America (USWA) was recognized in Youngstown, the CIO Council organized its own Labor Day celebration. In 1946 more than ten thousand workers attended a Labor Day picnic at Whippoorwill Grove. The day featured foot races, baseball games, boxing matches, dances, art contests and guest speakers.[43] All the workers I interviewed had consistently participated in union sponsored Labor Day picnics and shared, along with John McGarry, the feeling that "Labor Day at Idora Park was a day of fun."

The function of the union for the steelworkers was not limited strictly to that of workplace representation or institutional opposition to capital. The union was an agent of social formation. While it

would be stretching the analysis to call most postwar local steelworkers' unions "community unions," there was never a complete
separation between the workplace and the community.

Some locals, such as U.S. Steel Ohio Works 1330, had numerous
links with the cultural aspects of their members' lives, and those
links continued well into the 1970s. The local often demonstrated
the capacity unions had for bridging separate, competing identities.
In 1948 the union refused to support the Community Chest charitable fund because some of its affiliates refused to give employment to
blacks. A decade later Local 1330 took a courageous stand against
racial intolerance by sponsoring a community civil rights meeting to
protest lynchings in the south. In fact, Local 1330's efforts consistently appear to have been a direct attempt to shape class formation
and consciousness by encouraging workers and their families to get
involved in local sponsored activities as an alternative to participating in those organized by either company or non-working-class institutions. For example, Youngstown steel companies sponsored basketball teams in the city's Industrial A League. Normally, younger
union workers would enthusiastically sign up to play, but in this
case the union resisted because the games were played at the
Youngstown Central YMCA. The YMCA had a nice, modern gym
and was heavily used by the community. The facility was the perfect
setting for working-class basketball—except for the fact that it refused entrance to black workers. In response, Carnegie Steel (later
U.S. Steel) organized a Jim Crow league at the West Federal Street
YMCA. But Local 1330 countered by organizing its own integrated
team and appealed to the membership to boycott company-sponsored
activities.[44]

Union athletic teams for softball, bowling, and golf leagues
brought thousands of workers together outside the plant gate. In
most instances, athletic events were just extensions of the work shift
without the bosses. Many workers, like Charley Petrunak, simply
changed clothes and places but never had to break away from their
co-workers. According to Petrunak, "The bowling league at Holiday
Bowl started at 11:00 P.M., and we bowled after working night turn.
We bowled until 6:00 A.M. Just the guys from the mills. About sixty
of us went all the time. We used up half the lanes. You could also
continue bowling all morning long for a buck."

Bowling leagues were usually organized by plant department. In
one YS&T league, eight departments competed against each other in

Steelworkers and good friends: *third from right,* my father; *second from left,* Bill Mastrangelo; *third from left,* Kenny Wanamaker, and *fourth from left,* Lou Veltre spent long hours socializing at neighborhood bowling alleys.

year-round competition. While some teams could include foremen, most squads were exclusively made up of a particular production and maintenance shop.[45] Steelworkers living in Struthers had their choice of three first-rate bowling alleys to choose from, and they bowled on union and shop teams entered in ten city leagues.[46] Within the Local 1330 CIO Bowling League, eighteen teams competed for "turkeys, hams and whiskey."[47]

Outdoors the union organized softball and baseball teams. Youngstown's premiere fast-pitch softball players competed in the city's AA League. Republic Steel employees and Local 1331 members fielded one of the better squads in the AA League.[48] Tom Kotasek was one of the workers on that squad: "While working in the mill I played in a Double AA Fast-Pitch Softball League at Oakland Park, and on a 1331 softball team in a union league. We played about three nights a week at Stambaugh Field and at Wilder Field in Warren. I played with the local team and Tiny's Bar in the Double AA League. We went to Labor Day weekend tournaments in Sharon. I would come home from Sharon at three in the morning and go to work at 5 A.M." Kotasek added that Local 1330 had a national championship team, because they apparently had the best pitchers. When time allowed, games were followed by "a lot of drinking and laughing."

Locals 1330, 1331, 1418, and 1462 represented the majority of Youngstown area steelworkers, and each sponsored city-affiliated Pony League baseball teams for thirteen- to fourteen-year-olds.[49] Union resources were available to members who sought assistance. Bob Dill explained about union meetings: "Guys would always bring up things that needed some financial support, like buying baseball uniforms, and the guys would always vote unanimously to support them." Baseball uniforms were an easy problem to fix.

When steelworking families strained to survive within household budgets, locals organized committees of housewives "to aid in [the] fight for food at prices that workers and their families [could] afford."[50] In addition, the steelworkers' locals in the valley administered a membership blood bank. Locals encouraged all members to fill out and return their blood-bank membership cards because, they said, "You never know when you or your family may require this worthy service." Failure to do so would violate the workers' "duty to protect [family]."[51]

The union was also of course a source of welfare relief during long strikes. During the 116-day industry strike in 1959 strike funds and food vouchers assisted thousands of workers. The support provided by the union often meant the difference between autonomy and welfare. Community standing was partly dependent upon a worker being able to preserve his home and family, and the union helped Oscar Flemming do this. During the 1959 strike, Flemming said "[I] got food vouchers from the union during those days to get by. My first one was for $76. That bought a lot of food then. Because of the union I never had to go on public assistance."

After the shift whistle had blown, union stewardship continued in the community. The line between community relations and workplace identity was often blurred by the union itself. Under headings like "Ye Gossip Column," union papers printed, among other inter-personal items, the dating practices of select single workers, workers doing repair work and electrical wiring, tips for managing money, wedding announcements, graduations of children, births, and funeral notices. [52]

Russ Baxter was a five-term president of YS&T Local 1462. When I asked him if he socialized with friends from the mill, he described how a union officer's relationship with the rank-and-file operated at both work and home: "I visited friends many, many times. I talked about a lot of social problems in the homes of other guys. One Puerto Rican worker was being screwed by a foreman who was discriminating against him, and that was hurting his ability to provide for his family. In fact, I spent a lot of hours in the homes of workers."

John McGarry was a union leader who fought grievances, led local strikes, and represented the membership at conventions. As president of YS&T Local 2163, he sometimes administered a territory beyond the union hall, where he was responsible for a web of intricate personal relationships. McGarry explained, "[I] would visit guys' homes who were sick or injured, especially single guys who didn't have anybody and needed rides to the doctors or to the stores." Most of the time McGarry could help, but he said "If I couldn't do it for some reason, they would get angry at me, and one even said it was my job as a union official to take them." His institutional terrain was the workplace and the union hall. As a community resident he supported baseball teams, raised his family, and helped his neighbors. But when he nursed sick or injured union members and chauffeured single workers around as the union president, he was acting within the intimate relationships of class.

# 3 Fried Onions and Steel

Steel mills are enormous medieval-looking structures with beefy smokestacks reaching up into the sky. In Youngstown, they dominated the landscape. All the eye could see was an industrial plantation that rivaled any nineteenth-century southern cotton holding. Within each of these metal fortresses was a complex layered maze of catacombs, furnaces, grinding cranes, finishing shops, and a cornucopia of steel products. Always moving through these spaces were thousands of workers. They started with a pile of iron ore and through a routine resembling alchemy turned out steel. As it produced steel, the mill was forging the class-consciousness of its workers, and this chapter focuses on the development of shop floor culture.

Mill life, according to my father, involved more than just making steel. The relationships among workers also entered the mill with the workers and found another expression inside. All the workers I spoke to noted that decades of steelmaking could turn people into mere fixtures of the process. If the workers were not careful, the daily grind would consume their spirit and turn them into exactly what the company believed they were, "costs of production." To counter the psychologically restrictive nature of their work, steelworkers spontaneously recreated their workspace, interjecting their community life into the work process. This linked their class status at work with their class relations at home. It was not so much that home ever reminded a worker of the shop floor, but that at times the mill could bear a striking resemblance to the neighborhood.

## Born at Work

A steel mill is a scary place, a dangerous place to make a living and produce a good product. Every worker knew the dangers of mill work. A review of some of the jobs involved in steelmaking reveals the risk that many workers assumed. The dirtiest and least desirable place to work was the coke plant. Workers cooked coal dust in ovens in order to produce coke, an essential ingredient in the making of steel. Despite only a brief stint in the coke works, Tom Bergman never forgot what it was like:

> The first time I worked there I think I lasted three hours. Man, it was hot down there. The fumes, the heat, and everything else were terrible. They would push out an oven every ten minutes. They would use coal dust, and they poured it into an oven which had holes on the top of it. They would bake the coal dust with gas heat. It would bake for eighteen hours; then a pusher would shove all the coke out into a small buggy. The coke would then be cooled by a million gallons of water. I left work that day and said I didn't care if I didn't get paid. Lots of summer help came to work in the coke plant. They always had a big turnover there.

The job of "oven dispatcher" exposed a worker to "extreme heat while working on batteries" and to "extreme conditions of dust and fumes."[1] I remember how opposed my father was to my brother Rick's decision to work summer relief in the coke works. Rick lasted less than a week.

While unskilled jobs were likely to be done under bad conditions, highly rated positions did not ensure comfortable, safe surroundings. One of the highest-rated jobs in the open hearth department was "ladle craneman." The job consisted of operating an overhead traveling crane that transported molten steel for ingots.[2] The job was rated Class 16, and despite its importance to a major producing unit, craneman Bob Dill said, "[I] just wanted the check and to get home safe and sound every day because my job was very dangerous."

Having a job of relative importance to the operations of each department did not protect a worker from dangerous conditions. The maintenance department was crucial to the continuous functioning of mill equipment. Downtime meant lost money. One of the key jobs within the mechanical maintenance division was that of "millwright." My father spent the latter part of his career as a millwright.

Tom Bergman did different jobs at the mill. On all the jobs the workers were treated like "animals" by the company. On many occasions Bergman and his union brothers walked off the job to protest abusive working conditions.

His primary responsibilities were to "inspect, repair, replace, install, adjust, and maintain all mechanical equipment in a major producing unit or assigned area." Despite having a Class 14 rating, the job was done in surroundings described in the following way: "Continual dirty and greasy work. Usually wet. Occasionally exposed to hot materials. Could fall from equipment while making repairs. Climbing in and around operating equipment."[3]

In spite of its dangers, the mill was also a place to socialize. While workers rarely talked about their job routines at home, they incessantly interjected their home and community life into their work relations. They talked to each other "in the canteen, on the job; it could happen anywhere and at any time."[4] Tony Modarelli was typical of many workers who found time on the shop floor to pursue social relations:

We didn't get no breaks, but we made our own coffee time. And we used to get in a group before the shift and talk about different things. Or after the shift, we would sit and wait for the clock and talk about our life and joke around. You know, I was always early to the job. One kid would make coffee and we'd get there and bullshit a little bit before starting at 8:10. We used to talk about our wives, stupid things done at home; some criticized things at home, others bragged about kids, and you talked about things that were happening in the mill. Talk was pretty open and we socialized a lot in there.

Modarelli and his wife, Viola, had prepared for our conversation by jotting notes on a small piece of paper. Most of his notations were about the functional aspects of his job. He wanted to be sure to accurately describe what he did at work. But when I asked him about interacting with other workers on the job, he put the paper aside and with a smile began to rattle off stories. His anecdotes revealed two provocative characteristics of mill life. First, workers were in a hurry to get to the plant because it was a place where social bonds were created and expanded. Second, work-related issues and community life were not segregated. Many of the workers I interviewed admitted to enjoying their job. Time of arrival was apparently one indicator of a fondness for shop interaction, and my uncle Frank said, "[I] couldn't wait to get in that mill. I would get there so early on seven to eleven that I had to punch an eleven to seven time card."

According to my father the desire to be at work was not just to converse with other guys: "Guys would play dice, football pools, and the 'bug' [illegal numbers]. We would throw ringers on the finishing floor and in the open yard during breaks and breakdowns. A lot of this happened on night shift because there were fewer supervisors. It all made the job fun. This way it wasn't so bad to go to work the next day."

Dad spent roughly the first sixteen of his mill years stenciling and cutting pipe in the electrical weld's finishing department. For the last fifteen years he worked as a millwright, repairing machinery in the butt and continuous ("B&C") weld department. In the intervening years, his work time was invested in the shipping department, where he primarily bundled and hooked batches of pipe for transportation on railroad cars and semitrailers.

Handling pipe was an intense and potentially dangerous job. Anywhere from six to eight pieces of pipe, varying from 6 and 5/8 inches to 16 inches in diameter, had to be tightly wrapped in three steel strands of wire. A crane would then lower a cable, which the "hooker" would have to carefully connect to the middle strand holding the pipe together. The crane operator would then move the pipe to a flatbed. However, before the pipe could be placed on the flatbed, workers would have to construct a temporary floor of wood four-by-fours or two-by-fours, strategically spaced on the flatbed. Wood columns would also be staked in each corner of the flatbed to create a frame for the pipe being shipped. This structure made it easier to pile the pipe and to remove the crane cable without destabilizing the load. After the first wood floor was covered with pipe, the entire process was repeated until the flatbed was bulging with clanging metal.

My father liked the people in the shipping department, especially the men in the electric weld and B&C departments, but he decided to bid on a crane operator's job in the finishing department. He was awarded the bid and for a short time operated crane on the finishing floor and in the pipe yard. "You could make more money running crane," but there was a down side to the job. The crane was a lonely place. Dad stayed on the crane for about four years and then went into the B&C weld department. Mom felt that Dad went back into the weld because he "missed his buddies." He apparently gave up a chance to make better money for a job which afforded him greater fraternity. He admitted that he "wasn't that hungry" and was willing to sacrifice dollars for company.

The plant was a social microcosm. Some workers went so far as to nearly recreate the neighborhood inside their respective shops. Anthony Delisio's shed was a good example of how a formal work area could be transformed into something altogether different: "I did a lot of cooking in the mill. Guys would bring wild game in. The company gave me three lockers, and I turned one into a kitchen. Guys would bring their food to my shed, and I had hot plates and skillets. Guys came from all over the mill. People would smell the garlic and onions from all over. We gathered around the canteen and then usually ate at the shed. I think the guys looked out for me because of those onions!"

I wondered just how Delisio found the time to do this at work. Most steelworkers did not have an official lunch period, and they usually ate next to their station. Jim DeChellis explained that "there wasn't much time to just talk." Consequently, workers depended on each other to create the breaks they needed out of a technical arrangement that made it all but impossible to do anything but work.

"The [finishing] mill ran all the time and we didn't have a lunch break," DeChellis pointed out, "so you ate right next to your job." The coiler operator process was very intricate, and DeChellis said, "if your buddy couldn't give you a hand and take over, you were stuck solid for eight hours."

Time to eat, cook, or throw ringers had to be stolen from production space. Anthony Delisio said that his job gave him time "between grindings." Other workers became adept at using the production floor to market goods. Delisio recalled in astonishment, "This guy would go around and buy things and bring the stuff into the mill to be raffled off, and we once raffled off an old car."

Sometimes, time to play would arise not from production but instead of production. As James Rich explained, "[We] had a lot of fun at work during breakdowns." Halts in production gave workers a chance to "talk about sports, the family, religion, and play cards in the mill." Rich achieved shop floor notoriety with the football pools he used to run. "Big winner always took the pot," he claimed. "We got up to 150 guys playing from the strip mill, the open hearth; people came from everywhere." This betting was so much a part of shop culture that they "even posted the weekly winners on the bulletin board next to the time clock."

Despite a dangerous and demanding physical environment, workers had ample opportunities to turn parts of the plant into places where they could satisfy their social needs. Breakdowns occurred fairly often. Tom Kraynak worked at Youngstown Sheet and Tube (YS&T) and kept detailed reports of his workday. His records showed that in one ten-month period, 16 percent of scheduled work time was lost due to equipment breakdowns.[5]

Porfirio Esparro may have found the most entertaining way to use the shop floor. Esparro was a talented man who finished high school by attending adult classes at night, before going to work on the graveyard shift, 11 P.M. to 7 A.M.. While employed with YS&T, he headlined a local band with one of the men from the mill, and they played in clubs and on WBBW radio on Thursday nights. Esparro played at the Starlight and the Colony, and his daughter, who is also a singer, was going to record one of his songs, "Yo te Quiero Mucho," which means "I love you so much." Where did he find the time to practice? "My friend and I used to rehearse in the mill."

Pranks and high jinxes were as common as steel coil. Al Campbell said, "Guys had their shoes painted orange when they fell asleep; and hot foots, too." Joe Flora remembered that workers would go to extreme lengths, disrupting the normal production day to get a laugh. It was not uncommon to see "water battles all the time," Flora said. "I've seen waste lit on fire on night turn, and it would smolder in a room. Damn near suffocated a guy." Augustine Sanchez laughed explosively as he recalled horseplay in the mill. As the perpetrator of a prank that involved a "guy named Kitchen" and a rat nestled between two slices of white bread, Sanchez was "almost killed." But cooler heads prevailed because "guys played jokes with one another all the time."

Anthony Delisio discovered one of many unique ways to use company equipment; it was called "pigeon blasting." Workers "used to shoot chalk out of air guns and crank that sucker up so that chalk would shoot out like a bullet." The goal was to try and hit pigeons resting in the rafters. Usually the pigeons escaped injury, but Delisio remembered it with a laugh. "When we were done it looked like an explosion on the ceiling!"

Workers more often knew each other by nicknames rather than by birth names. In most cases, the names reflected the work a person did, but they could also originate from distinguishing personal characteristics. George Papalko, for example, was called "Kayo." During his working days, Papalko had been a union activist and wondered whether the nickname had anything to do with his beating on "company goons." Others, he recalled, had more descriptive names. "A black guy who always had grease on his shirt, we called him 'Greasy Belly.'" Another man was "Goosie John," another "Flatfoot Floozy." Papalko assured me, "Guys got nicknames most of the time because you liked them."

Despite periods of boyish playing, workers treated the work space as a kind of sacred ground. They knew their jobs, needed the work, and committed their lives to making the world's finest steel. But it was precisely because they spent so much time working together for companies that most workers claimed really didn't appreciate them that they found ways to redefine the work area to make it meaningful to them. YS&T employee Arnette Mullins expressed a near extrasensory connection with his steelworking friends. As he put his hands up, palms facing one another, to frame his face, Mullins said, "Man, it was like a family, you spend thirty years with the same guys, you become real close, sometimes too close." Too close meant "Your feelings become his and his feelings become yours." Playful interaction at work generated the same intimacy and sense of inclusion that neighborhood visiting accomplished. In the mills, workers carried out the same social processes they encouraged in the community.[6]

## Inclusion, Exclusion

To a degree, the difference between the workplace and private life was transcended. But for steelworkers, there was a fundamental, unbridgeable dichotomy between those who lived by their labor and those who lived off the labor of others. Workers also drew distinctions according to the actual physical exercise of work, and these were not reserved for management. Workers were bitterly resentful of any worker who failed to pull his weight and as a result "burdened another guy."[7] When asked to describe "bad union members," workers unanimously used language very similar to that expressed by Tom Bergman: "There was always some 'free riders' on the union's coattails. They were suck asses and guys who dropped their burden on you. And they would use their friends in the company to protect them."

Bad union men were people who "didn't care about their fellow worker," and "seemed only interested in themselves."[8] If there was an ironclad rule among workers it was "you didn't stick other guys."[9] Instead of helping one another, "bad ones served their own purposes."[10] In contrast to men who did not put their fellow workers first, Bergman recalled a previous union president who made the ultimate sacrifice for his men: "[Chuck] Kabal led a wildcat in 1952 that led to the company firing the whole union executive board, but he made a deal with the company to keep him fired but rehire the rest. He gave himself up for the rest of the guys." Bergman also mentioned the greatest sin against class—being a company man. Workers were intolerant of men who "were too quick to side with the company" or did not have what it took to stand up to the company.

Beating the company appeared to be a significant criteria for judging union mettle. The less "bullshit" a worker took from the bosses, the more he was respected. While workers were not dogmatic about being right all the time, they were adamant that their peers not treat them as if they were wrong.[11] There were good and bad workers, but even a bad worker wanted to beat the company.

According to Republic Steel employee Oscar Flemming, the "best union man" was a fellow black worker named Pete Starks. In fact,

John Varga took great pride in his work and in his membership in the USWA. Here, in Youngstown's 1953 sesquicentennial parade, he stands next to a plaster model of a Bessemer converter mounted on a float sponsored by Local 1330 to honor the role of organized labor.

Starks was considered exceptional by a number of Republic workers, white as well as black, because "he knew the union front and back," "knew what it took to win," and further, "the company was intimidated by him . . . so he got things done."[12] In order then to protect worker interests and to trump the company, according to Flemming, "Guys would come from all over the mill to get Pete."

The value placed on class solidarity and opposition to the companies' agents regularly stimulated plant-wide group discipline. Under normal conditions group cohesion was maintained by spontaneous agreement, union insistence, or peer isolation. But occasionally, as John Varga explains, more coercive means were used to ensure class loyalty:

> We once refused to work a sixth day until everyone in the shop was in the union. At first, it wasn't a closed shop. One guy didn't want to join but he wanted to have the same benefits. He demanded certain jobs, but unless you're a union man you can't demand anything. We stayed

out for one turn. This guy thought he had rights. But he had no rights outside the union. Yet he stayed out for about a year and we let it go. But then he began to bad mouth the union. To get certain jobs you needed the union's protection, and this guy wasn't union, so he didn't get protected. He couldn't get anything on his own. The guy just didn't want to pay those few bucks but wanted the overtime and six days and all the rest that came with being in the union. So five of us went to the boss and said, "We'll come out on the sixth day, but we aren't going to do anything. The quicker you straighten this out and stop giving that guy six days, the sooner we'll get back to work." So they sent him home. He joined the union next week. We were 99 percent union, and one bad apple wasn't going to ruin the bunch. So we had to step on his toes. We had some real die-hards who didn't tolerate any bad mouthing of the union. There were guys who had their tires slashed because they weren't supporting the union enough. Guys would think twice about shooting their mouth off about things they knew nothing about if their tires got cut. This guy was calling the union corrupt and a bunch of brown-nosers, while he was getting the benefits of the union. You shouldn't be running the union down.

Except for the war years and for a slight change in the 1970s, women were a rare presence on the shop floor. In 1979, an estimated 6 percent of the workforce at YS&T was female, and most of them were assigned to one plant. The Campbell, Struthers, and Brier Hill plants of YS&T employed approximately 251 women, but only 20 of them were at Brier Hill. In addition, in the three plants combined only 4 women worked in craft positions. The records of Local 2163 further reveal that in 1973 only 20 women out of 2,270 workers were dues-paying members of the union.[13]

While male workers seemed to have accepted these women, there is little evidence that men ever saw the mill experience as gender neutral. Women were generally considered "real good workers," but most men remembered very little about them. Oddly, Arthur Newell's wife, Marcia, claimed, "The only time I heard him [Arthur] talk about his job was when a woman welder came into the mill, and she could cuss better than any man!"

What the workers recalled about women in the mill was that some were attractive, some very strong, some needed the money, and that most male workers were not adverse to helping women workmates. Comments were made by a number of men about their memories of women in the plant, but there was little difference in what was said.

None said anything at all about collective grievances, feelings, experiences, or actions. Mary Ross admitted that while the men were very nice to the women, they never included them in any "union-type business." Given this apparently masculine orientation, feelings of being unwanted were undoubtedly shared by female workers. One such woman, Elma Jones Beatty, recalled that female workers "felt the mill was a place for men to work." Beatty worked at Republic Steel from 1940 to 1944 and considered the mill a "dirty place—a man's world."

Race issues were more contentious than gender differences. Race often strained, even if it never broke, the class dimensions of industrial production. Before the passage of civil rights legislation in the mid-1960s, most mill departments were segregated by job. But beginning in the late 1950s and accelerating through the 1960s, the companies gradually dismantled the separation and job segregation that characterized the early days of steel production.[14] As YS&T employee John Barbero explained to labor activists Staughton and Alice Lynd, "In the open hearth we didn't work successfully to end discrimination until the Kennedy years. . . . somehow, the same people who harassed the blacks in the Truman and Eisenhower years, under Kennedy their sense was to do the decent thing, accept it, and not struggle at all. Our department was desegregated and blacks moved into all the jobs. I didn't hear any complaints at all."[15]

It was not until spring 1974 when the Federal Northern District Court of Alabama approved an affirmative action agreement between the steel companies and the United Steelworkers of America (USWA) that job ghettoization stopped.[16] In parallel motion, union policies and practices also began to better reflect the needs of their minority membership. In past cases of white workers who resisted working next to a black man, the union too often went along with blatantly discriminatory company policies. But in the early 1970s successful efforts at inclusive union politics began to make a difference. One impressive example occurred in 1973, when black workers at Local 1462 entered into a coalition with "progressive" white workers behind the union leadership of Ed Mann.[17]

Before the changes in union policies, white workers predominated in the shaping mills and on most craft jobs. Black and Hispanic workers were concentrated in the coke works, cinder plant, and blast furnace. These parts of the mill were traditionally thought of as outposts for unwanted jobs. YS&T worker Kenneth Andrews explained,

"The dirty, lowest paid jobs, which were just cleanup, were black jobs. . . . every now and then a black guy got something better. For the most part as a black worker you got the worst jobs." Jobs that were all-white, like "masons, machinist, pipefitters, were kind of elitist."[18]

Job occupancy also appeared to be associated with ethnicity. Many Italians were masons, Irishmen tended to be railroaders, Hungarians and Slovakians congregated around the open hearth, and foremen were usually "Johnny Bull" English.[19] Before the late 1960s work assignments tended to tribalize the working class into different departments. But the department where a worker was assigned did not necessarily determine where that worker traveled. Youngstown's steel mills were monstrous structures that covered miles of property and housed thousands of workers. While job sequences could be rationed, it was practically impossible to close off physical areas by race or ethnicity. There was no denying that workers got around the plant.[20] Most of them knew in fine detail the specific shape and operations of areas other than their own. After all, most experienced workers had performed a multitude of diverse jobs.

Most craft workers were not restricted to one part of the plant. White masons, millwrights, and electricians would go anywhere a breakdown was reported. In addition, along with a white mason went three or four black bricklayer helpers. Thus, the plant was opened up because of the need for skilled workers to work everywhere and the need for laborers to always be on site. Most large jobs also necessitated a setup crew, and as a result, it was very common for unskilled laborers of different nationalities and races to be preparing an area for skilled workers of primarily one ethnic group.[21]

Mill property was never conducive to common amenities, and the steel masters were hardly charitable in providing inviting lunchrooms for casual lounging. While that made eating more inconvenient for workers, it also prevented any chance of white-only dining areas. As it was, workers normally ate at their stations with whomever was on the job. Perhaps a group of workers could walk away from the job to eat, but as my father said about racial congregating in the mill, "Where could you go to get away?" It was not that black and white workers didn't willfully separate, but that separation rarely added up to exclusion.[22]

The most important exclusion steelworkers experienced was the physical and philosophical separation between themselves and their employers. A them-and-us mentality was dramatically evident in the concrete ways they responded to shop floor indignities. Workers, according to Charley Petrunak, were "good people":

> We were poor and growing slowly together. We were experiencing good and bad times together. I mean, one day I saw a guy eating a sandwich, another changing a light bulb, and another standing close by pissing in the corner. The next day, guys were starting to feel like people instead of being treated like animals. Guys would be afraid to speak up to the boss about a dangerous condition. But then with the union getting stronger we started to fix these things. Built sheds and piss houses. We took care of shit. You made steel together and you won your dignity together. You couldn't make steel alone and you couldn't win respect alone.

Petrunak punctuated his remarks by using his whole body to make a point, and the disregard for conventional decency suggested by his colorful description is precisely the kind of company behavior that generated a practical sense of working-class identity.

Augustine Sanchez identified with people who worked for a living, saying, "they need one another to live a good life." This need was born out of the company's treatment of the workers:

> Once we shut down [workers walked out] because we didn't have a fan to cool the job down. We asked the foreman for over a week to get a fan. They kept promising but never delivered, so one day I came in there and there was no fan so I sat on my fanny. The bar came down, and it just kept going. The head roller came over and said, "Why didn't you catch it?" I told him why, and he sat down and started to chew some tobacco. The general foreman came to me and I said, "No work without a fan!" Then another supervisor came down and chewed me out. I said, "Look at how I look! I'm sweating like a pig!" And I'll tell you the rest of the guys backed me, and eventually, later that day, the company found a fan. Then we started to work.

The demeaning actions of the bosses stimulated class solidarity. Tom Bergman vividly recalled that the indignity of company abuse was all about being treated as sub-human:

> I remember once in the early '50s they [the company] would take out the water cooler in the winter so the pipes wouldn't freeze, but they

didn't replace them in the summer. Well, it got real hot and we were on those furnaces, which were 180 degrees, but there was still no water cooler. So we said that's it and we walked. Well, they said, "No, don't leave, we're going to bring you some cold water in buckets." We said, "What do you think we are, cows and horses?"

To be a worker meant being subjected to employer abuse and also being treated "like a criminal."[23] Antigio Mosconi believed that worker unity was necessary "because of the abuse of the company" and that the union was an answer to those who "didn't live by God's law." Mosconi was a proud immigrant who took great offense at people who used the word "dago" to describe Italian-Americans. While he was in the Italian army, he had learned the value of ethnic and national respect, he felt that too often the company had been disrespectful. He poured me some homemade wine and recalled a personal experience of the company's disregard for fairness:

After a job one day, I never forget it was cold as hell, I went to get my card, and I was told my card was pulled until I see the superintendent at 8:00 A.M. They didn't tell me why. So I went to see Joe Carlini [Local 1331 vice president and grievance man]. Joe went to see the big boss and the foreman was there, and he said I was spending too much time at the coke jack [coal or wood burning in a barrel]. The general foreman said he couldn't do anything, so Joe grabbed the guy! I calmed Joe down, and we went to see the superintendent and convinced him to put me back to work. But they didn't pay me for the day they sent me home. So Joe went to the foreman and grabbed him and said, "Why didn't you pay the guy?" I got my money.

The workers' sense of common fraternity was also deeply felt as a difference in status. If workers did less well than their employers and had to fight for all they earned, it was, according to Joe Ryzner, because of class privileges. To him it was clear: "You got a different strata up there." After all, he said, "Some guys are always bosses, kings and dukes, and then you got down here the peons. . . . We were the peons, you know, the workers." A dichotomy between workers and nonworkers came down to "we worked, they didn't" and "they had money, we didn't."

Russ Baxter learned from his work experiences, stating "The bosses stood around and watched you while you worked." Before becoming a popular local union president, Baxter had been a professional fighter with 150 bouts to his credit. He was a deeply physical man who took great pride in his top-ten pugilist ranking as well as his YS&T work record. With a raspy voice, bulging biceps, and a decided limp, Baxter seemed a throwback to earlier rough-and-tumble union leaders. To him, real work meant using your body and getting your hands dirty. Workers were part of a separate class who were always "waiting for payday."

As workers explained it, they got less and managers and supervisors got more. This was commonly expressed by workers through the use of comparative metaphors such as John Occhipinti's, "One guy had a Cadillac and the other guy had a foot mobile." Mike Scocca also pointed out "[If] it wasn't for the union, we'd never have houses or cars, [and the company would] still be crapping on us."

## Risks

Accidents and injuries were a common aspect of mill life. Most of these were minor. But on occasion someone died, and while the company called it an accident, workers were not always so forgiving. Anthony Delisio recalled an event that burned a place in his consciousness: "One of the ugliest was when this guy walked over the catwalk, over the straightening rolls, and he got drilled by a pipe through the right kidney. It went in about a foot. It was a four and a half inch pipe, but because it was so hot it cauterized. He stayed like this for two hours with his insides partially hanging out. Guys spent a lot of time talking about it. It made you realize how vulnerable you were in the mill."

Regardless of what workers did inside the plant, many of them suffered a similar physical deficit from their work—a profound hearing loss. Joe Flora lost so much of his hearing that he now wears two hearing aids. Bill Calabrette and my father were among a number of retired Republic steelworkers who waited to confront their loss until their health coverage included hearing aids. Red Delquadri was another worker with two hearing aids who pointed out that when he was working as a "hooker" in the pipe yard he "didn't have ear plugs," and that "it was the constant pipe banging that may have done it." The noise was mind rattling, "a lot like a real loud clang!" Delquadri recalled that the "whole neighborhood could hear the bigger six-inch or eight-inch pipe falling into a pocket."

Hearing loss, as well as other workplace injuries, could function in an additional way to influence class solidarity. Safety issues often provided the spark for organized and spontaneous worker resistance to the company. Anthony Delisio, who was off work because of a hearing problem, was notified by YS&T that his failure to answer a "five day letter" to report back to work broke his continuous service with the company. Delisio responded by filing a grievance based on the claim that he "developed tinnitus from noise pollution" and that while the company provided earplugs,

> they didn't fit properly to mold to your ear canal. They had earmuffs, but they were nothing more than a plastic cover with a piece of foam rubber. Hearing protection was never provided, and my hearing deteriorated to the point where I had a ringing in my ear and suffered considerable hearing loss, about 70 percent on the high tones and 40 percent on the low. I hear about three different sounds constantly. The doctors told me I would be deaf in ten years if I didn't get out of the seamless.

Delisio's disability claim was supported by a doctor's report that "determined he has a severe bilateral sensorineural hearing loss." But the most critical part of the report was the doctor's finding that the "attenuation of his ear protection devices may not be sufficient to meet his physician's recommendation" (i.e., stop working in the seamless).[24] In light of the medical evaluation, the company decided to grant Delisio's request for a "Rule-of-65" retirement.[25]

On a very warm summer day in 1969, Local 1462 member Tony Pervetich, my mother's uncle, was working in the open hearth pit, cleaning out the dirt after a long run of heats. Pervetich had worked steady day turn for years and was seven days short of an announced retirement. A surprise retirement party had already been organized.

My mother knew little about her uncle's actual work, only that her steelworking father had said that the pits could be dangerous. Work in the pits was dirty, noisy, and, because of the dust, visibility was poor. The pit itself was small and box shaped. Workers would shovel waste material into the back of a large dump truck, which then transported the load to a dump site. While men shoveled debris, the driver had to navigate the truck in a tight restricted space. Local 1462 considered the pit unsafe because the vehicles did not have back-up signals. The union had filed a grievance over the issue and requested that a warning horn be installed "so when a truck was

My mother described her uncle Tony Pervetich as a warm, funny man who always put a smile on her face. His tragic death devastated her and sparked a
wildcat strike at YS&T's Brier Hill Works.

going to back up the people working there could hear it."[26] The company had rejected the grievance.

Ed Mann was the local's recording secretary at the time, and he
wrote, "Shortly after the grievance was rejected, a man who was
going to retire in about seven days was run over by one of these
trucks. He was crushed. He was a well-liked person who had worked
there a long time and was about to retire." Tony's death devastated
my mother and her father. Mom sadly recalled that after Tony's final
days in the mill "they had made plans to take a trip back to the place
of their birth in Dubrovnik, Yugoslavia."

Every worker witnessed shop floor injuries, and these injuries made them "feel like a piece of property," and as a result they "knew no one was indispensable."[27] Work was a contradiction that workers had to accept; it was the source of livelihood and the most threatening part of life.

The threat of injury, company abuse, and a sense of inclusion contributed to the workers' shared identity. Perhaps that explains why Tony Perry felt so humiliated when he retired. On Perry's last day at work, the company gave him a white hardhat with his badge number painted on the side. Forced "to wear the hat," he was paraded around the plant in a final tribute. So why was he humiliated? Only management employees wore white hats. Workers wore red, and when he retired he was celebrating a career as a worker.

# 4 Making "Good Money" on Time and Credit

Some workers took regular vacations. Niagara Falls was a popular destination. Many traveled to see relatives in other states. Over the course of a long career most workers saved up the means to occasionally get away. But not my family. We never went on vacation, and as my father recalls, he "never bought a new car." It was not for lack of interest. Dad had once enjoyed a trip to New York with the men from the mill, and Mom wanted to see more of the country. But how does one afford a vacation with four growing boys? For my mother, it always came down to a simple equation; "a dollar for play meant a dollar less for what was needed."

Material instability initially shaped my family's sense of "them" and "us," and it was a rare worker who believed in an economic equality between themselves and the mill owners. These differences in living standards were the economic origins of a steelworker's class consciousness, which this chapter traces.

## Them and Us

Consonant with theories of rising income and middle-class status is an attendant belief in the workers' contentment with capitalism. Conventional assessments have particularly focused on the workers' satisfaction with the wage-earning process. With industry-wide collective bargaining contracts, new high-skill technology, more group-oriented management practices, and the advent of managers who separated workplace control from ownership, many social scientists proclaimed that there was little ideological difference between workers and owners.[1] But all of what supported the myth of "happy wage" laborers should have been junked with the 1973 release of one major government study.

The "Report of a Special Task Force to the Secretary of Health, Education and Welfare" stated that "significant numbers of American workers are dissatisfied with the quality of their working lives."[2] Contrary to numerous postwar analyses that focused on, among other things, the alleged changes in class structure, the task force report examined the activities that work entailed. What they found was anything but encouraging to the prophets of a sedated, postindustrial American working class.

The study found that work in America was not fulfilling and that workers were not content with "relative satisfactions." Across the occupational spectrum, workers offered similarly trenchant criticisms about the work they performed: "dull, repetitive, seemingly meaningless tasks, offering little challenge or autonomy."[3] Instead of bettering conditions, the Cold War economic boom left work remarkably unchanged, and while pay was important, "pecuniary models of work motivation failed to address the human need for job satisfaction."

Arguments for the mediation of economic polarization were based on allegedly increasing social mobility and a creatively designed work process. But these went up in smoke when the task force reported that "Workers at all occupational levels feel locked-in, their mobility blocked, the opportunity to grow lacking in their jobs, challenge missing from their tasks."[4] While most Youngstown steelworkers expressed some genuine admiration for their employers, they were always clear about why they became steelworkers; "it paid good." Ultimately, the employers' inattention to everything but the profit-making capacity of their workers fostered a working class with serious grievances about conditions inside and outside the workplace.

No social indicator has been more uncritically correlated with class status than gross income levels. Conventional theories have argued that working-class identity after World War II was predicated on aggregate measures of lifestyle. The cumulative effect of these studies was to cast postwar industrial workers as middle-class consumers. But as Richard Parker has revealed in *The Myth of the Middle Class*, "distinct classes between rich and poor give a much more

realistic picture of America than does the notion of a single homogeneous middle class."[5]

Parker's work defies the myth of the 1950s' affluent society and de-
constructs the meaning of a middle-class life. Supported by official
government and business-oriented sources, Parker explains that
"[B]eing middle class can mean comfort bordering on opulence; but it
can also mean outright poverty, or deprivation that is only one step
removed from poverty."[6] The middle class was unofficially bounded
at the upper end at forty thousand dollars and at the lower end at four
thousand dollars. Absurd as this range was, social scientists in the
1950s and 1960s were so committed to the myth of the middle class
that they accommodated the wide diversity of income levels by creating something called an upper and a lower middle class.

The average postwar blue-collar worker was decidedly not in the
upper middle class. Parker estimated that the lower middle class,
which accounted for 70 percent of the population, received only 22
percent of the nation's total money income. He pointed out that the
"richest 10 percent of Americans in 1968 received more monetary income than the entire bottom half of the population."[7]

Inequality of income, however, was not the only problem with the
myth of a middle-class life style. Liquid assets (i.e., checking and savings accounts) were also important measures of a family's economic
stability. Parker stated that in 1969, "one-fifth of the population
owned no liquid assets whatsoever, and nearly half the population
had less than $500. Less than a third had more than $2,000." In addition, among the lower middle class the average wealth of a family
(i.e., the sum of its assets minus debts) was a paltry $6,000.[8] Parker
clearly showed the fallacy of assuming the existence of a broadly homogeneous American middle class. For the lower-middle-class, blue-
collar worker, domestic life was characterized by heavy debt purchasing and consequently "danger [was] never far away."[9]

Andrew Levison further demonstrated in Working-Class Majority
that "The majority of America's blue-collar workers live not only far
below affluence, but below the modest government standards for a
fully comfortable middle American life."[10] By using the Bureau of
Labor Statistics' "standard budgets" for 1970, Levison pointed out
that the majority of working-class people were living below the government's "intermediate budget" of $10,670. Approximately 60 per-
cent of all working-class families were living below the "middle
American standard of comfort and security," and placement of a ma-

jority of workers "midway between affluence and poverty" was sheer propaganda.[11]

According to Levison, in 1970, 60 percent of America should have been identified as working class. He insisted that differences between manual and intellectual work, and labor that separates "clock punchers" from professionals and executives, represented a fundamental occupational divide.[12] Through a comparison of social, economic, and working conditions, Levison repeatedly showed how working-class life diverges from a middle-class living. It was his contention that "workers and the middle class are divided by a real and profound inequality," and that for purposes of class consciousness this inequality is not abstract but "a visible daily reality."[13]

A "suburban" steelworker's domestic budget provided very little evidence to erase the dichotomies of class. National studies of steelworkers' hourly earnings before 1950 indicate a substantial climb in gross wages from 1940 to 1945.[14] The growth rate, however, appears large because steelworkers' wages had a long way to climb. At the outset of World War II, annual wages were so suppressed that, as unbelievable as it may seem, "most steelworkers were earning Depression-level real hourly wages on the eve of Pearl Harbor."[15] Yet wage increases after the war still left steelworkers with less disposable income than was needed. In 1952, the federal government's Bureau of Labor Statistics estimated that a "modest but adequate budget" for a family of four was $4,132, or $79.46 per week. The average steelworker, however, earned only $78.30 a week.[16] In addition, according to the Consumer Price Index, in percentage terms price increases had risen twice as high as wages.[17] It was only in the last third of the following decade that worker incomes began to genuinely improve. Annual earnings, adjusted for inflation, did rise by 2.6 percent per year from 1967 to 1979, but so did income tax payments.[18] In the mid-1970s, a "good" year's earnings would amount to approximately $21,000, but total taxes could eat up roughly 28 to 30 percent of earnings.[19] While the lives of steelworkers demonstrate an unquestionable economic improvement from the standard experienced by their parents, they also reveal the slender threads that held up the appearance of a middle-class life style.

Historically, Youngstown steelworkers have not always earned good wages. None of the workers I interviewed said they made any "good money" until the late 1960s or early 1970s. George Porrazzo remembered, "In the 1960s the electric weld was a bad place to work.

It was bad because we worked four days a week, and I had to labor, I had to hook up, and I was laid off so much I ran out of unemployment one year." To make matters worse, "at that time [he] had two kids and unemployment was thirty-nine dollars a week and the sub-fund was about fifty cents." Truth was that "all through the '60s and the early '70s I was in serious debt." Debt would have been easy to fall into in years like 1967 when Porrazzo was laid off for fifty-one days, or in 1970 when he experienced forty-six "payless" days.[20] It was not until the 1970s that workers who had been in the mills since the outbreak of World War II started to feel a modest level of comfort: "Until then you didn't make any money and your benefit package didn't amount to anything."

With better pay came larger durable purchases. Home buying in the late 1950s stimulated an increase in consumer needs and spending. But the comforts of modern conveniences, modestly etched into a worker's home, amounted to an edifice of good faith. For Youngstown steelworkers, middle class consumption was done on credit and time.

Red Delquadri worked his entire career as a semi-skilled "hooker"; he always intended to leave this line of work. He half joked that his problems started at Ellis Island when the immigration recorder misspelled his last name. But once he was in the mill, it didn't matter how his name was spelled, because he was called "Red" all the time. Delquadri and his wife, Mary, spoke with me on their front porch. He noted that the only way he could afford home furniture was to not only borrow and buy on credit but to purchase the display samples. He usually maintained a running balance on loan accounts and once his debts were paid (not a single record of missed payments), immediately began to accrue new ones. Outstanding yearly balances from the late 1940s through the '50s of one hundred dollars or more were not uncommon on loan accounts.[21]

In addition to keeping up monthly payments on lending balances, workers also paid for their improved standard of living in installments. Domestic staples, like meat, milk, bread, toiletries, and appliances were often part of a line of credit at the neighborhood store.[22] My uncle Frank's wife, Annie Frattaroli, worked and shopped at a corner store a few feet from her home. She recalled just how rare it was for steelworkers' families to pay for items at the time of purchase. As economic conditions oscillated between better and worse, consumers "could easily get credit at the independent stores, because they knew their patrons and it wasn't odd to have no one pay." People "came in from Powersway and Cameron to buy their groceries and put it all on credit." All that the proprietors ever required from the buyer was to "just to let them know your situation."

The personal relationships that local merchants had with their working-class customers were not only a normal shopping convenience but, during hard times, a lifeline for household budgets. In 1959 the steelworkers struck the industry for 116 days, and for many workers it was the credit extended by area merchants that ensured the solidarity of the strikers. Bob Dill, Frank Frattaroli, Cayetano Caban and my mother pointed out the consideration workers were shown. Dill said, "What really helped us all around here was these little ma and pop stores. My milkman would give me five quarts of milk and he never thought about collecting money. I shopped at Toriello's and when the strike was over I owed him eight hundred dollars. In fact, no one would send you a bill. When the strike was over you took care of your bills." Frattaroli added, "When we went on strike I went down to Strouss and asked about buying things, and they looked at my record and said I could keep buying anything I wanted and they wouldn't send me a bill till one month after the strike was settled." Caban remembered, "I had credit at an Italian grocery store, Buchille's, and those people were wonderful to me. Whenever I needed food they gave me credit. I loved those people like they were my brothers." And my mother had a story:

One woman at the meat store in New Middletown helped us out. I went one day to buy some meats, you know, the cheaper cuts, and she's talking to me. Nothing real important, just this and that. And I'm pointing to the meat I want, and she takes the big, expensive cut and wraps it! Doesn't say a thing. I didn't know what to say. She knows I can't afford it. But she keeps saying, "What else?" That night I came home with all kinds of meat. She knew I was having problems. I don't know how she knew, she just did.

Even skilled workers with steady employment relied on borrowing to attain the trappings of middle-class America. During his working career, Joe Flora bought the following "big ticket" household items on credit: compact sweeper, bedroom suite, hot water tank, front door, table saw, dining-room furniture, chandelier, carpeting, electric

range, refrigerator, freezer, washer, dishwasher, color television, and
stereo.

Buying on credit was a slow creep toward middle-class respectabil-
ity that was often interrupted by the vagaries of production. Workers
bought the items they needed to live more comfortably one expenditure at a time. Joe Flora's consumption pattern reveals a conscious
decision to never get too far ahead of his income potential. Consider
the following purchasing sequence for some of the items listed
above: freezer (1966), washer (1967), color television (1968), and electric range (1971). Despite an improving gross pay, Flora usually limited his significant expenditures to one a year and never exceeded
two in any calendar year.[23] Workers were buying some of things
middle-class consumers bought, but they were buying more slowly
and cautiously.

This strategy was a common and necessary practice for steelworking families who wanted to live more conveniently. Every purchase
had to be planned and traded off against things not purchased, because earnings were variable factors. Despite the better days eventually realized, a steelworker's yearly wages did not automatically rise
with the average per capita income of Americans. Armando Rucci,
for example, averaged $183 a pay period in 1970. But in 1950 he actu-
ally averaged $242. Rucci's financial surge, however, generated at the
dawning of America's "suburban decade" was brief; in 1951 and 1952
his paychecks usually amounted to less than $100.[24] Joe Flora's an-
nual mill earnings never functioned as an indicator of a constantly
rising standard of living. Listed here is the direction his income took
beginning in the "better days" of the 1960s and ending the year
Youngstown Sheet and Tube (YS&T) announced it was closing down
operations: 1963–66, up; 1967, down; 1968–69, up; 1970–71, down;
1972–74, up; 1975, down; 1976, up; 1977–78, down.[25] Wages could
even dramatically fluctuate over a few months. In 1968, Armando
Rucci's earnings fell by almost 60 percent in six months.[26] Steelworkers' earnings, therefore, could wobble enough to loosen the tenuous grip they had on the next rung of the economic ladder.

Perhaps no economic characteristic better defined steelworkers
than the roller-coaster activity of their savings accounts. Workers
who managed to put some money away usually soaked it up during
long layoffs or strikes. Many workers told me that during the 1959
strike they lived off their savings. For example, John Costello answered the question "How did you cope during the 1959 strike?" by

Joe Flora worked for three different local steel companies and volunteered plenty of his time to do odd jobs for St. Nicholas Parish.

saying, "Well, I had some money saved up for the upstairs of our house, so I spent that. I had three kids to take care of." Brad Ramsbottom answered the same question by saying, "I was renting at the time with three kids and was saving money to buy a house. We didn't buy one until 1962. I used most of that up during the strike."

Anthony Delquadri's financial history illustrates how difficult it was for steelworkers to invest and save their way into the middle class. Delquadri began 1947 with $856.37 saved up from his military service. After one year back on the job his account was down to $654.50. By October 1950, he had only $10.77 in savings. His balance dwindled to $2.62 in the summer of 1962 before climbing in December of that year to $105.05. But by May 1965 it had fallen to $10.93 and did not rise much until a $300 deposit on 11 August 1970. It was not until Delquadri's last year of active employment that he was able to save more than a thousand dollars. In fact, it was only after he retired from the mill and took a job for two years with a local medical facility that he managed to save nearly $6,000.[27]

## Unsteady Wages

The struggle to save money was made particularly difficult by the irregular work available for many steelworkers. No matter how good wages were, if a worker was not employed for most of the year, earnings would shrink and debt would rise. My father's biggest obstacle in trying to support a family as a mill hand was the on-again, off-again nature of work. He estimated that from the late 1960s to the late 1970s he was "out of work a quarter of the time." A normal work pattern consisted of being "off four or five weeks, then [he]'d go back for a short time, then get laid off again." Steady work seemed to come in waves, but inevitably "things got bad and [he]'d go a couple of weeks without working, then put a few days in and get laid off again."[28]

Over the course of a long career, layoff time could amount to a considerable loss of earning potential. Jim Visingardi followed his father's footsteps and spent the biggest part of his life working in the mills. He willingly offered information about his working experiences, and this was informative because he had kept close tabs on his lost time: "There were many weeks with three or four days, and there were countless layoffs and times they sent you home because this broke down or that. I would never say that mill work was steady.

I sat down and figured out that I actually had five years accumulated layoff time. Would you believe that? I was once laid off for sixteen weeks in a row. In a row! I'm not bullshitting. Now this is not counting strikes or walkouts. I'm talking time when they tell you they don't need you!"

While the American economy was experiencing a Cold War boom, Charles Freeman encountered only bust. As he explains, not only were the 1950s not prosperous times, but the 1970s were only slightly better: "During the 1950s I don't think I worked a whole year. In fact, in 1957 I was laid off the whole year. Sometimes four or eight months at a time. And in the 1960s it was just a little better. My best year was about $14,000 in the '70s. I survived though. . . . I hustled for any work I could find."

Perhaps one of the more unforeseen consequences of accumulated layoff time was that workers found it difficult not to think of themselves as working class. Working one day and then off the next solidified the belief that what workers did was a job and, unlike the bosses' work, not a career. While their tenure at the mills was secure, their time on the job was erratic and confounding. Earnings bobbed up and down from year to year, like apples in a barrel of water, and the sense of social mobility that comes with middle-class status could emerge only from a comparison with the previous generation. Brad Ramsbottom was one of many workers who believed that he had not received fair compensation for his labor, yet he took some satisfaction in knowing, as he said, "We got much more than what our parents got and that mattered a lot."

A steelworker's working-class identity was also determined by his ability to earn beyond the base hourly rate. Earnings, for example, could drop by a third without incentive pay added on to the base. The bonus portion of a worker's total earnings could range over a three month period from 28 percent to 47 percent.[29] In 1974, George Porrazzo, for example, took home a gross pay in which the "regular" hourly earning fluctuated between 51 and 61 percent of the total.[30] Incentive pay clearly had a dramatic effect on a steelworker's standard of living. There was one major problem with this source of compensation: it was not money a worker could count on.

Not all jobs were rated for incentive, and the rates could fluctuate. In addition, premium pay and cost-of-living adjustments were contingent on market conditions and contract language. For most workers, incentive pay was a feast or famine situation. Armando Rucci never earned a nickel in extra pay from 1950 until the mid-1960s, and once he began to earn the money it was not a dependable source of income. Rucci's first pay check in May 1968 was 23 percent incentive earnings. But his Thanksgiving check for that year only included $4.80 in shift differential on top of base pay.[31]

Base pay and opportunities to earn extra pay were linked to a worker's job class. This placed a premium on the job sequence a worker was locked into. The importance of gaining access to flexible job ladders is best seen by examining the segregation of black workers into the lowest-paying departments. Before the opening of apprentice programs in the 1960s, black workers were in job ghettos like the coke plant, the blast furnace, and the cinder plant. The coke plant of YS&T's Campbell Works comprised three divisions: batteries, coal handling, and by-products. Within the batteries the job sequence ran from a Class 5 "lidman" to a Class 12 "pusher operator," and there were seven class steps. But within the coal-handling unit there were only five steps, and the top job was a Class 10. Thus, workers assigned to coal handling and unable to bid into other units could get trapped into a lower job sequence that would determine their lifetime earnings.[32]

Black workers as a group were also more likely to be assigned to labor and helper jobs than were white workers. Until the post–Civil Rights period, higher paid bricklayers in YS&T's mason department were nearly all white, while black workers in the unit were always bricklayer helpers.[33] In addition, despite the sizable role that helpers played in completing a masonry job, incentive pay was completely controlled by the bricklayers. Boyd Ware and Charley Harp were black helpers who never felt they were paid what they were worth. When asked about receiving a fair wage they sternly replied, "No, no way!" The injustice grew out of the fact that they "had so much to do and the bricklayer only had to do the job." Helpers "prepared it, helped with it, and cleaned it up," and they "were the first on the job and the last to leave." Yet according to Ware and Harp, "The bricklayers were paid one-third more! Only the laborers made less money then we did." Pay differentials on jobs of similar nature, difficulty, and skill requirements could corrode worker cohesion.

Kenneth Andrews was a helper in the mason department at YS&T. He was sprucing up the yard for an upcoming Fourth of July picnic and interrupted his work to talk with me. He had a lot to say about different pay arrangements. He was not only bitter about getting "the dirty, lowest paid jobs" but complained, "White guys would be shoveling upstairs and I'm shoveling downstairs and the heat is just as bad, but this guy is making fifty dollars a day and if I was making twenty dollars a day I was lucky. And we were doing the same job!"

Wide income disparities, restricted job sequences, fluctuating incentive rates, lack of access to highly rated jobs, and irregular bonus pay were not only unstable sources of middle-class mobility, but also weakened a worker's sense of group attachment. "The truth is," according to ex-Local 1331 president Joe Carlini, "that the incentive system was the ruination of the workingman."[34] Workers knew that incentive rates were used by the companies to squeeze more production out of the labor force. They also knew that if left unchecked, incentive plans would increase competition between individual workers and units. George Porrazzo was all too familiar with how bonus money undermined class cohesion:

> Incentives caused problems in many ways. Guys would fight over who could work the jobs that made eight hours bonus in eight hours. Now you might make four hours bonus, but you tallied 1,800 pieces of pipe on one shift. You didn't have time to take a leak! The shippers were hot all the time over incentives because it was supposed to be split over all the mills. So if a mill went down the guys in the working mills would be able to split the incentive bonus two ways instead of three.

Production meant pay, and faster production under a complex incentive system meant better pay. It's not surprising that some workers "had a habit of really risking [their] life to keep production up."[35] But trying to make a little more money not only could be bad for your health, it could also damage worker unity. If the conditions of work could spell out a "them" for workers, they could also confuse the issue of who the "us" was. The tenuousness of incentive pay needs to be understood in the context of a worker's overall earnings. While steelworkers were the highest-paid manufactory workers in Youngstown, they were desperately dependent on bonus money.[36] If rate pay and overtime were essential to a worker's ability to provide the things his family needed, then he could not just walk away from what was being offered. Instead, workers found a way to control the arrangement that was designed to manipulate them, and consequently they demonstrated a renewed commitment to class fraternity.

Steelworkers knew that piece rates were bad for workers in general but that incentive systems could raise average take-home pay. Merlin Luce was employed at U.S. Steel and wrote for the *Ohio Works*

*Organizer.* When asked if incentive programs created conflict, he replied, "Yes, but a very tough and impressive act of collective action, a wildcat, was held in our department over a straight incentive system. So, here you have the workers advocating something we were theoretically opposed to, because we needed some way of squeezing more money out of the company." This explained how radical means could be used for seemingly conservative ends.

The workers manifested the "inner laws of wage labour" by acting collectively on behalf of a "friendly" rate system.[37] How did they do this? Their most common strategy was to assure that rates were set artificially low and to police recalcitrant rate-busters. Ironically, union contracts aided in this covert endeavor. The contract spelled out guidelines for when new rates could be set and old ones changed.[38] Consequently, workers knew when and how the rates were going to be adjusted. "We worked out a system on incentives where when the clock was on nobody did more than, say, thirty-five pieces," revealed Armando Rucci. "But," he continued, "when the rate got set and the clock was off, we did what we knew we could, which was forty or forty-five." Corrupting the employer-imposed rate required that workers get "together and figure out what [they] needed."

The tentativeness of the workers' rising standard of living can be further confirmed by the fact that many steelworkers held second jobs. If workers were becoming middle-class Americans, then steel-working jobs should have been enough to secure social mobility. But despite the grueling physical and mental demands of a forty-hour mill workweek, the vast majority of steelworkers periodically held down second jobs. Even excluding strike periods, some modest amount of most workers' pre- and post-mill shift time was spent working elsewhere for a wage.

Workers held second jobs because they did not believe they could improve their living conditions by relying on a steel company check. Jim Visingardi went to a job after he had punched out at the mill because, he said, "If you lived off your mill salary you were really restricted in what you could make." The bottom line was, he continued, "You couldn't depend on the job," and workers needed to remember that a "four-day week was a 20 percent cut!" While second jobs were not always easy to come by, steelworkers were familiar with other manufacturing industries.

Charley Petrunak, for example, worked for two years with the Pennsylvania Railroad while employed by YS&T. Tony Modarelli, Al Campbell and Willie Floyd laid concrete, and Armando Rucci did landscaping jobs. James Rich tended bar and worked with an electrician on his off days. He held second jobs because, he said, "The income from my mill job was insufficient; I wanted my family to have more." Joe Flora worked at YS&T from 1957 to 1981, and every year earned additional money by doing from one to three other jobs.[39]

Paul Dubos is a soft-spoken man who worked at YS&T and enjoyed woodworking. Although he suffered from curvature of the spine, he "worked with a carpenter building houses." He used vacation time as well as time after morning mill turns to build other peoples' houses. All in all, he "would do about thirty-five hours a week." He eventually became such a good carpenter that he built his own house.

The steelworkers' unwillingness to rely on their steel earnings contradicts the notion of a "$26-an-hour" manufactory employee fully comfortable with the benefits of a modern free-enterprise system.[40] A more accurate portrayal of postwar working-class conditions was that the present looked a lot like the past. After long hours in the mills and tremendous profits for the steel companies, workers reaped such modest rewards that many were required to work during their vacations and often to add another fifteen to thirty-five hours a week in side jobs to attain something like a middle-class standard of living.

## The Costs of Resistance

While class awareness cannot be directly inferred from the workers' economic conditions, their individual and collective actions can have a powerful influence in creating such an awareness. Strike activity had relevant implications for the development of class identity. Six national steel strikes occurred from the end of World War II until the first shutdown in 1977 of YS&T's Campbell Works.[41] These strikes meant loss of pay and increased hardship. During the periods mill operations were shut down, workers were forced to resort to an assortment of means to survive. This surely erased any lingering doubt that they were different from the mill bosses and owners.

During the 116-day national strike in 1959 the majority of steelworkers relied on personal savings, side jobs, and family support to
survive. Both the union and the company anticipated a long strike,
and some workers had accumulated a lot of overtime early in the
year. But with nearly twenty-six thousand steelworkers involved in
the forced shutdown, many were forced to rely on other creative and
institutional forms of support. During the first week of the strike
fourteen hundred workers filed for public assistance at the Mahoning
County Welfare Office.[42] By the midpoint of the work stoppage the
welfare office had distributed $659,589 worth of assistance to steel-
workers. Included in that amount was $7,785 for 81,313 pounds of
federal surplus food.[43] Just 60 days into what was to be a 116-day
strike, relief rolls had increased by 87 percent.[44]

Along with county welfare assistance, however, the United Steelworkers of America (USWA) provided vouchers and the names of
local merchants who would accept them at their face value. Despite
having a fair amount of strike resources, workers' material needs
grew so acute that the USWA District 26 Counseling Service in
Youngstown reviewed twenty-five thousand relief claims.[45] Even
making allowances for some reporting errors and reoccurring claims,
a sizable portion of Youngstown's highest-paid manufactory workers
were forced to act as if they were indigent. At the end of the strike,
steelworkers whose only previous relationship to charitable community agencies was as donors had received $1,300,000 in support from
all public relief sources.[46]

Whether or not a steelworker applied for public help, he or she was
still forced to be creative to make ends meet. Kenneth Andrews, a
bricklayer's helper at YS&T, was "hired to drive a truck a couple of
days a week for one company and washed cars for a guy who had a
car lot." Tom Bergman was a Republic Steel employee and Korean
War squadron leader who found himself in a real bind as the strike
began:

> We moved into this house on July 3 and we went on strike on July 14.
> Family was here and they helped us, and we paid our own medical in-
> surance because the company canceled it while we were on strike. We
> also just paid the interest on the house, and the payment came to $54 a
> month. . . . We didn't have any sub [supplemental pay] in '59, and I got
> one $30 strike fund check from the union. We did get some surplus
> government food. I also worked some for the old man DeBartolo [a

prominent Youngstown commercial real-estate developer], cutting his grass and watering his lawn for ten, twelve, hours a week for $1.60 an hour. I kept doing this during layoffs because we didn't always get un-employment.

It is hard to imagine how Andrews and Bergman could feel that they were a part of just another economic interest group when in order not to lose a home they were taking care of other people's property. Unfortunately, not all the workers could keep their homes. In spring of 1959, my family purchased a home in a barely developed rural area. We moved out from my father's parents' home in Youngstown, where my mother and father had lived since they were married. After only a few weeks of the strike, my parents, concerned that they could not afford the mortgage payments if the labor struggle persisted into the fall, decided to sell the house and move back with my grandparents. Losing the house was a disappointment and a sign that economic advancement had been forestalled. But according to my mother, what came next was a lesson in personal fortitude: "We used to pick tomatoes and can vegetables. We would pick a monstrous bushel of tomatoes and get twenty-three thick quarts out of it for a dollar. We went out to a farm off of Route 62 in Canfield. One time we were hurting, and we heard about the Amish digging potatoes for one farmer. So we followed the Amish and whatever they left behind we picked up and took home."

In 1959, Youngstown was a heavily urbanized city, but in adjacent counties and just across the border into Pennsylvania there was a considerable amount of farming. During strikes and slow periods, black and white steelworkers were recruited as day farmers. Charley Harp got only one food voucher from the welfare office for eighty-seven dollars, but recalled how fortunate he was most mornings to find space on a farm wagon. "Most of the time I picked potatoes and beans," recalled Harp. He would wait downtown every morning at 6:00 A.M. for "spot labor" to come by with their truck looking for a few strong backs. Harp scoured those outlying farms because work was work and he was married with three kids.

Red Delquadri found himself in the incongruent position of providing personal service for one wealthy capitalist: "The '59 strike was a long one. Money was going out but nothing was coming in. It dragged on, and to make a little money I went back to the country club to caddie. I was in my mid-thirties. And William Pollack of Pollack

My parents enjoyed plenty of good times attending dances at the Idora Park
Ballroom. But during the many difficult periods they worked the Amish potato
farms to put food on the table for four growing boys.

Steel used me quite a bit and paid me five dollars for nine holes. That
was better than trying to get help from Catholic Charity. They
worked on a most needy basis, and for Christ sake I had three kids
and they said I wasn't one of the worst cases!"

Steelworkers who lost homes, delayed buying homes, picked
beans, gleaned after Amish farmers, went on welfare, and carried the
golf clubs of the elite were not embittered Americans. They were,
however, aware that they suffered unjustly at the hands of the steel
firms and that their bosses never had the need to "bum cigarettes in
the street."[47] The consequences of doing what family survival required were connected in the minds of workers to the structural lim-
itations of their jobs.

While workers did not call what happened to them "exploitation,"
they knew that their disadvantages were an element of workplace relations. Workers also knew that they were forced to struggle while

company management was treated as a privileged class. No one expressed this thought more forcibly then John Occhipinti. Occhipinti answered my questions about company treatment by shooting out every word. He was angered and said, "The foreman had steady work, and if we were on strike or there was a big layoff, they would send the foreman to school for training; they didn't give a damn about us."

If anything, the experience of fighting to hold on to hard-earned symbols of success and prosperity increased worker cohesion. It was, as Arnette Mullins stated, "the struggle that brought you together." This was the basis for the us-and-them mentality. Workers knew from the cut of their clothes to the cut of their holiday meat that they were different from their steel masters. When added to the daily inequities on the shop floor, difference became opposition. As that happened, them-and-us changed to them-*versus*-us.

# 5  How to Steal a Wheelbarrow

In the Youngstown mills, the steelworkers' us-and-them attitude was frequently expressed as shop floor resistance. *Us* and *them* were discrete classes, and they existed in opposition. But this attitude equivocated, and it varied across the ranks of the workers. This chapter examines the forms of shop floor resistance that expressed worker resentments.

I interviewed workers like George Papalko and John Pallay, who recited a litany of company abuses and believed that "class war was constant." They had lived through the Steel Workers Organizing Committee's (SWOC) organizational drives in the 1930s. Both had heard Gus Hall, a Youngstown Communist Party member and labor organizer, speak on socialism.[1] Hall, according to Papalko, "was a good influence" who "emphasized that [they] were part of history and should be oriented towards the poor man around [them]." Pallay revealed that his father-in-law was a "Socialist who worked in a meatpacking plant in Akron, Ohio, who always kept the *Daily Worker* around a lot."

For workers like Papalko and Pallay there was a struggle going on between "two classes—us and the company." Yet despite living through a violent organizational period and having been exposed to leftist ideological influences, neither ever believed that the mill or the steel produced belonged to the workers. A little communist preaching did not change this: "We were just concerned about when payday was."[2] Merlin Luce would have agreed that some communist influence would have been helpful. As a long-time member of the Socialist Workers Party, Luce was inspired by a socialist vision of work that underscored the importance of class resistance.

The majority of the steelworkers I talked with, however, remembered earlier periods when "work was good" and the "bosses cared about making quality steel." To many it was a source of pride to be making steel in Youngstown. But workers like Porfirio Esparro believed that after 1969, when Youngstown Sheet and Tube (YS&T) was sold to "out of towners," the new owners treated the workers like a "cash cow to be sucked dry." Others, such as Tony Pellota, stressed the difference between the old days' "foremen from the ranks" and the modern "guys right out of business schools."

Every worker I spoke to understood that his employer treated him as a member of a lower class, and as a group the workers recognized a multitude of class inequalities. Every day they struggled in mundane ways against their employers' efforts to "push production out." But struggle coexisted with "good bosses," an obligation to put in a good day's work, and a respect for profit making. What conditions caused opposition to flare up into resistance? When the steel companies resisted union demands or asked for unacceptable contract concessions, workers went on strike. When a foreman acted in an abusive way, workers turned the shop floor into a battleground. But when workers were allowed to do their jobs without undue interference, or when "the money made you comfortable," then peace reigned. Antagonisms, it seems, were driven by the situation and the interests at stake.

Class is what "defines people as workers" and no concept of workers' consciousness could credibly ignore the ideological effects of production.[3] While consciousness of class fermented at various levels wherever workers interacted with each other, the workplace was essential to class antagonism. It was on the job that workers accumulated their grievances against the other class, the class that does not work. Although absent from the neighborhoods, the boss was always present in the workplace, and the shop floor introduced workers to a potentially common antagonist. The mill was fertile ground for emerging hostilities, and workers' resistance to their employers' authority was expressed in a mélange of coercive confrontations and subtle, indirect assaults.

## Memory and Resistance

Youngstown steelworkers struck the steel companies six times between 1946 and 1962. In the 1950s alone, Youngstown steelworkers
totaled ten months of downtime in official strike action. Excluding
the usual three months of layoffs every year, the average mill worker
was withholding his labor for 13 percent of the scheduled workdays.
This may underestimate strike time because during the deep reces-
sions in 1951 and 1957 layoffs were a regular feature of life for many
Youngstown steelworkers.

Despite the number of official strikes, the story of resistance by
Youngstown steelworkers against their employers is more intricate
than the occurrence of national work stoppages would suggest. What
kind of resistance, for example, characterized the noncontract-year
strikes? According to the workers I interviewed, worker resistance
never ceased, and it reached a level of defiant anarchy in the anti-
Communist McCarthy years. Contracts and consumer purchases not
withstanding, in the period from 1950 to 1977 workers did not hesi-
tate to "shut down" a job, a shop, or an entire department over something they felt was unjust.

In contrast with official strike action, every case of wildcatting
(i.e., an unauthorized strike) was loaded with worker venom. Surprisingly, most of the workers I interviewed reported that during official
strike periods no deep-seated antagonisms toward the company were
ever manifested. The one major exception to this attitude was the violent 1937 strike. In 1937, Youngstown steelworkers engaged in a
monumental battle against the YS&T and Republic Steel Companies
that resulted in the shooting deaths of two workers and gunshot injuries to at least twelve others.[4]

Tradition played an important part in formulating worker resistance. Most workers in Youngstown had no formal educational history of unionism in the valley or of the labor movement in America.
Most did not have an extensive knowledge of their local union's history (although all knew who their president was and most had a fair
grasp of their contract), district union elections, or for that matter
the leadership of the International Union.[5] But nearly all could recite
accounts of the blatant oppression by company bosses that necessitated a union-organizing drive. With few exceptions, workers were
inclined to take our discussions back to the "early days" before the
union.

Curiously, although only 4 percent of the workers in my study started working in the preunion period (before 1942), every one of them remembered or knew it explicitly. Where did their knowledge come from? Some of course came from the men who had experienced the days of boss autocracy. Tony Nocera, now eighty, said, "[I have] seen things you wouldn't believe." His still-strong Italian accent announced that he was one of the old-timers, and he knew more labor history than any other worker I met. Nocera stressed that organizing was necessary, if decidedly dangerous: "It was hard because union activity could get you fired, and they did fire people. We did organizing on the side, around the furnaces; you know, away from any supervision."

But to organize, the workers had to get inside the plant. The process of getting hired was humiliating, but because it was crucial to survival, workers submitted to the bosses' will. Nocera experienced a preunion practice that epitomized the workers' powerlessness:

In the old days, foremen would pick the guys they wanted from a lineup outside the plant gate. You would pack a lunch and go out for the eight to four shift. If you didn't get picked, you went back out at four with your lunch, which was now your dinner. If you still didn't get chosen, you came back out at midnight. Your lunch was now your breakfast. While day after day I didn't get picked. One day I didn't go out on night shift. The next morning the foreman asked me where I was the night before. I told him, "You never pick me so I didn't come out." He said, "I needed lots of men, you should have been here." What could you say, you had no power and you needed the job.

In later conflicts, the memory of being "cherry picked" by the foremen would surface in the form of a current indignity. Whatever the particular issues were at the time, workers would invoke the historical image of their own and their fathers' begging for work. Workers often called upon a "remembered natural link between labor and its product, and in the name of this memory" resisted the contemporary efforts of steel masters to limit its meaningfulness.[6] Little wonder, then, that workers raised to sacred status "the [union] adoption of a schedule, because [they] knew [they] had a job and [they] could count on it."[7]

Working may have been the end and class solidarity the means, but becoming a mill hand only opened a worker to a demeaning array of daily denigrations. "If you got lucky enough to be picked you had to do whatever they told you," stated George Papalko, remembering the power of the bosses in the early days. Workers had little room for mistakes and were denied any right to question authority. Papalko said that if a worker hesitated or complained, the bosses would yell, "Take your goddamn bucket and go home!" Workers could be replaced indiscriminately and for arbitrary cause, like smoking a cigarette. But more often it was for signing up with the union. The result was, according to John Pallay, "Somebody else got your job."

Workers paid a heavy price for the union cause. At a time when "the bosses were like gods," being part of a union meant salvation. George Papalko remembered not only the danger but the power that came with fighting the gods on behalf of a union:

> Gus Hall and guys like us would break into scab houses and wreck them, just tear them up. We'd look for the company stooges. During the 1937 strike the company had gunmen in the "number 1" crane in the shipping department. They killed one guy there. Shot him dead. And the police were on their [the company's] side. I belonged to the "Flying Squad." We went to the Warren Plant where the company was flying piper cubs over the plant to drop food to the scabs. The planes flew so low you could hit them with stones. They did the same thing over the Market Street bridge in Youngstown. That's where they would try to bring scabs into the plant. There were two roads that ran under the bridge. They had two new Ford stake-body trucks that they got from Donnel Ford, with a machine gun mounted on the cabin. They would come crashing right through the pickets. So we decided to put cars across the entrance and roadway. The Youngstown police brought a master cruiser, with shotguns under the roof, to pull the cars out, so we put telephone poles across the road. Scabs would even swim across the Mahoning River to get into the plant.

Tony Nocera recalled that the strike was over union recognition and the workers' demands for a contract: "We went out in '37 to get a contract. It was bitter, tough. The governor and the county sheriff sent troops against the men. The National Guard was called in, not to help the workers, but to help people who wanted to go back to work. It was a bitter strike. Strikers were even killed. It was actually the National Guard that broke the strike."

Nocera's terse account reveals three notions that I believe steelworkers fed off to marshal later shop floor resistance: the reason for

the bloodshed was a contract, military force was used against workers, and scabs tried to take union jobs. The themes are repeated in
Red Delquadri's account of 1937: "There was shooting going on and
scabs in the mill. The National Guard was there in the street to keep
the workers from going after the scabs. They were there working for
the company. But we had to have a contract."

With or without independent verification, workers believed the
worst of the company. James Rich recalled that "the building across
the bridge on the Campbell side of the mill had an encampment of
machine guns," and that "the company had dynamited the bridge in
case the workers tried to storm the plant." Marvin Wienstock had
heard that in "the church steeple at the bottom of Market Street
Bridge, Republic Steel had stationed 50 caliber machine guns." Carl
Beck was a participant in the strike as it was carried out against Republic Steel, and he cited the LaFollete Committee Investigations to
confirm the company's murderous acts: "Two pickets were shot
from machine guns, which were mounted on cranes in the old tube
mill. The angle of the bullets and their trajectory had all come from
the elevation, the height of the cranes in the old tube mill."

What happened in 1937 became a reason for each future worker's
dedication to unionism. The mill owners' transgressions during the
strike were later transformed into offenses of a contemporary nature
and served to continue the demonization of the company. Workers
gave me accounts they believed to be true of mill bosses who would
leave the plant during the midnight shift and "go up the hill to sleep
with the wife of a worker." This, they said, was the cost of keeping a
job. George Papalko and John Pallay swore this indignity to be true of
the old "hunkies who lived up the hill" from the Stop 5 entrance of
Republic Steel. They claimed it was common for workers to bring
homemade wine and fresh vegetables into the plant as a bribe to the
boss in exchange for another day's labor. Much later, John Varga was
fond of the way his "coffee brought foremen" into his crane for a
friendly visit. Once it began, their conversation sometimes turned
toward items that department workers needed. In return for a favor
done for the boss, workers' schedules were actually written by Varga.
He never told the men that he was responsible for scheduling, because "they were happy and so was the boss." Varga personally bene-
fited by receiving jackets and other small tokens of appreciation.

Nearly all of the workers made the intriguing observation that
today's unions were struggling because they forgot where their

strength came from. As one worker said, "Young guys today act like the company gave us things we needed because we worked hard and needed them."[8] The anti-union era was so compelling that later workers would hold each other's behavior up to the badge of that "terrific fight . . . to help bring the union into this Valley."[9]

By the time of my interviews, the events of more than a half century past were clothed in intrigue and had been elevated to a class mythology. The strike story has been told elsewhere, and eyewitness accounts have dramatized the actual sequence of actions.[10] It seems to me that what really happened is not as important as how workers' beliefs shaped their class attitudes.

For many, the knowledge of the fight for a contract, the violence, and the strikebreakers was handed down from a father, an older friend, or a close relative who had worked in the mills. Brad Ramsbottom was inspired to union stewardship because he remembered his father, "who worked in the mill in 1925 telling [him] about guys who would go the mill everyday, on all three shifts, and never get work." Ramsbottom was vice president for Local 2163 and along with his dad, they accumulated seventy-eight years of service at YS&T. Jim Visingardi remembered his "dad bringing a lunch to the mill and having to come home and then going back at 3:00 P.M. and back again at eleven." Watching his dad do this day after day was all he needed to become "a good, strong union man."

But transmission of this history was not necessarily from father to son. Surprisingly few of the men I interviewed spoke directly to their fathers about mill work. Despite the relative silence of their fathers, sons knew what the old days were like because they overheard family and mill conversations and remembered that, according to my dad, "men were killed down there at Stop 5." For ten years, my father worked at Republic Steel while his father worked at YS&T. The Little Steel Strike occurred when Dad was eight years old. His father and Uncle Louie were involved, but they never told him about what happened. Dad does, however, have a strong memory of the day the shooting broke out: "We were right there, up the hill from Stop 5 [main entrance to the Republic Steel plant, off of Poland Avenue]. You could smell the tear gas and hear the commotion and the gunshots. There was lots of activity, people coming in the house and running out. And I remember a lot of crying." My mom too recalled how at age seven she saw her mom crying and heard from someone in the house that her father was out "fighting the company goons." His defense of unionism eventually lead to his being branded a labor agitator and fired by YS&T.

## Wildcats

The 1937 strike was exceptional. Strikes from 1942 to 1966 for improved contractual items never raised the level of class bitterness that the Memorial Day walkout over union recognition released. But wildcat strikes, an interesting divergence from the official, top-down, approved union action, were characteristic of the postwar years. Not only were spontaneous walkouts proof of ignited them-versus-us resentments, but they also suggested that employer relations were a constant source of worker solidarity. The actual depth of worker opposition fermenting inside production may have been an intriguing revelation for Ed Mann. As he recalled, "When you get people out [wildcatting], sometimes you can't get them back. Their response is 'So what else do we want?'"

Despite the great respect workers had for the collective bargaining contract, they could wildcat with abandon. John Zumrick boasted, "Even after the contract was signed we still pulled strikes. . . . and in the boilershop the contract didn't matter, we would strike any time." Zumrick seemed a man capable of just such apparent insolence. His square face, sharp chin, and snow white, billowy hair conveyed a no-nonsense attitude about things. His favorite recollections were those about "working when [they] wanted to, and telling the boss where to go when [they] didn't."

Most wildcats lasted only a short time, but some lasted for an entire shift, and they could persist for two or three days. Augustine Sanchez explained that shutdowns of "thirty minutes or so" were very common in the rolling mill at Republic Steel. He took pride in how the action "would cost the company a lot of money." Zumrick's wife, Rose, recalled that "a lot of times they would pull the shop out when the company brought in outsiders to do construction and lay off some of the guys. So what they'd [the workers] do is pull the shop out for two or three days."

The spontaneous collective act of bringing production to a halt was a working-class technique to force the steel companies to change an unjust behavior. The practice of wildcatting, however, was not simply a reaction to a temporary shop floor power imbalance. To many workers, walking off the job was a way to govern their own labor time. Jim Morris relished the opportunity to explain how workers exerted their control: "This was a pretty strong union shop. The guys remembered the days before the union, and they weren't going to go backwards. If the company didn't abide by the contract you were out the door. That was the way we controlled them. They were so used to ruling that the only way to get any say-so was to shut the place down. They came to heel when we shut the operations."

Workers apparently were so bold about stopping production that they became a source of embarrassment for the local union leadership. Consider the following exchange I had with Local 1331 president Joe Carlini and Local 1331 benefits officer George Porrazzo:

> R. Bruno: Were there wildcats?
>
> J. Carlini: I can remember that in the 1950s we struck every week over all kinds of things. It was too hot; it was too cold; every time a guy wanted to go home he would come up with something he didn't like!
>
> R. Bruno: Are you saying workers walked off the job just to avoid working?
>
> G. Porrazzo: All the time. Let me give you an example. One day we walked out at the beginning of the shift, and when we got outside we found out that we had walked because the water cooler in number one weld wasn't cold. That was chicken shit.

To some workers, wildcats had gotten so out of hand that the act of withholding labor began to lose its effectiveness. Stopping work was often necessary, but they "couldn't go out for just anything." Workers had to have a solid objective. The problem was that one man's or one unit's solid objective was another's "chicken shit." John Druzoff, for example, did not think cold water was a wildcat issue, but he thought the incentive system was a valid reason for stopping production: "I think they [Truscon Steel] had a screwed-up incentive system. Someplaces made good incentives and others couldn't make any money. So many guys couldn't make any bonus, but guys doing almost identical work were making a lot more money. We used to go on strike nearly every other week over the incentive rates. We also used to walk out over guys who were fired. Usually it was over the incentive rate that a guy got into trouble."

Then again, Tom Bergman proudly recalled that once, because the company would not hook up a cold water cooler, they "left the plant and went across the street to the beer gardens and cooled [them]selves."

While workers could disagree over the importance of water coolers, they found common ground over job loss and safety matters. The workers in the blooming mill of Youngstown Sheet and Tube's Campbell Works were a good example of shop floor militancy aroused by job loss. In 1975 the entire department refused to report for work because eighteen men had been eliminated due to a new incentive program. The walkout idled the mill's 5,800 workers for three days. The death of Brier Hill open hearth worker Tony Pervetich (described in Chapter 3) sparked a walkout by the day-turn crew and the afternoon shift; that walkout was eventually joined by the men on night turn. By the next morning, the entire open hearth was shut down: "Every area—pit, cranes, floor—was represented."[11]

Ed Mann believed that the wildcat was important in shaping class consciousness: "Other departments didn't go out on strike in sympathy, but there was just no work for them. We made the steel. That cost the company a lot of profits. Everything cost, measured as trainloads of material coming in or scrap half loaded. That's a feeling of power. And it isn't something you're doing as an individual. You're doing it as a group."

In expressing their opposition to their employers' terms for work, workers found numerous ways to subvert the production process. Regardless of what plant supervision or the "bible" (i.e., the contract) directed, workers "did the job the way [they] wanted to." The conditions upon which that job got done were determined by the workers, as my father explained about scheduling and job assignments:

> One foreman would never let you leave the job before the shift ended to wash up. But guys left. We would cover up for those who wanted to leave earlier. You made your time up the next day by coming in early for the guy who helped you out. Sometimes guys left early before the other guys showed up, and the whistle would blow announcing a breakdown and millwrights were supposed to answer that call. But no one was there to go. Now some millwrights would be there, but the company insisted on assigning all of them before the shift started. Guys could have gone on the job and fixed the problem, but they didn't move until they were told where the foreman wanted them.

If workers could, but did not, respond to breakdown whistles because they were abiding by company policy, who was controlling the job?

## Grievances

It was, of course, not uncommon for workers to genuinely cooperate with shop floor directives, but when they did the job without complaint it was because they considered the procedure fair. When they thought the work process was unjust or not in accordance with the contract, "Guys would get really angry and scream and yell at the boss."[12] After screaming, they would file grievance upon grievance. This opposition becomes surprisingly meaningful in light of how workers used the contract mechanism most similar to judicial practice. Most progressive critics of modern collective bargaining have pointed to the grievance procedure as a government-imposed device to preserve the unobstructed flow of capital.[13]

While the litigation process has obvious flaws, it is the workers' willingness to use the contract provisions that suggests a conscious reaching for methods of resistance. Perhaps more important than when or how often grievances were filed was the workers' direct participation in the process. It is plausible to assume that if workers were more concerned with preserving a friendly, nonhostile relationship with their boss they would insist that their union committeeman fight the grievances on their behalf. On the other hand, if workers felt it was more important to stand up to the boss than it was to maintain "a same team philosophy," they would attend grievance meetings. Merlin Luce felt that workers always had to make the decision between direct confrontation or relying on the union as a "legal representative that works for the men." Luce peppered his comments about functionally handling shop floor disputes with a captivating social analysis of capitalism. Surrounded in a den by numerous books written by Trotsky, Lenin, and Marx, he argued that in order for workers to effectively confront their employers, they needed an understanding of politics and economics. Without such a study, however, too many workers would unfortunately chose a misguided "team" approach.[14]

Under United Steelworkers of America (USWA) contracts, grievances at all levels could be formally presented without the actual grievant in attendance. However, as a matter of union practice and politics, grievants usually appeared at arbitration hearings. According to the workers I interviewed and the limited documents available, it appears that most workers were eager to fight their own battles. Brad Ramsbottom made the interesting point that the "biggest gripe about the union was it was not militant enough [on grievances]." When asked whether workers went into their own grievances, he recalled that "most guys went loaded for bear and raised all kind of hell." On one occasion, Ramsbottom chuckled, "We had to restrain one guy from beating up a company official, Dave Haney, right there in the grievance hearing!"

The presence of a worker at a grievance meeting or arbitration hearing depended on the type of problem. Where the union filed a charge of contract violation over company interpretations that affected a wide number of workers, the union was usually represented by an attorney or district staff man and the local president.[15] In certain wage and working-condition cases, however, workers went so far as to write their own notices to the company, detailing not only their point of disagreement, but their lack of trust in others to solve shop floor problems. One such case involved seven YS&T pipe repairmen who took exception to not being considered "craftsmen" under an addendum to the contract. They wrote "To who it may concern, . . . we highly resent the company's attitude" toward trade and craft jobs. The letter then asked, "Who in the company echelon has the right or the duty to consider some as craftsmen and reject others that are doing the same work?" But to make the point that they were not expecting the union to speak for them, the workers then asked, "Or, perhaps, is it someone in the union echelon that is responsible for this?"[16]

In cases of individual discharge and discipline the meetings typically included the grievant. Undoubtedly union officers had institutional reasons for wanting grievants to attend these hearings, and workers could seriously jeopardize their cases by not appearing. Still, attendance afforded workers a chance to sit, not as a subordinate but as an equal, before a neutral party.

The formalities of this legal proceeding gave workers the opportunity to do three things they could not spontaneously do inside the plant without fear of retribution: defend themselves, accuse the company of being unfair, and demand justice from their employers. One dramatic example of a grievant challenging the company involved a worker discharged for "using profanity towards a foreman, threatening the foreman with physical violence, pushing the foreman, and

conducting himself in a belligerent manner." In this case the worker
not only successfully convinced the arbitrator that there was "serious conflict in the testimony" over what actually happened, but
made a more persuasive argument then the company's attorney, the
superintendent of Industrial Relations, the Industrial Relations coun-
selor, and the assistant superintendent of Industrial Relations.[17] I
suspect this had to be a very invigorating experience for the rank-
and-filer.

Not all grievances were filed by individual workers on behalf of
self-interest. On many occasions, a group of workers would collectively charge the company with violations of the contract, not only
to protect themselves but to save the jobs of their peers. Examples of
group grievances include the following: "13 grievants protest reducing the number of Head Tablemen . . . 80 grievants protest reducing
the number of Head crane Followers . . . 29 grievants protest the
company not scheduling a Clean-Up Man . . . 21 grievants protest as-
signing Transfer Car operation to Crane Follower . . . 89 grievants
protest the reduction of Threader Operator and Die setter jobs."[18]
While in most cases this action made practical sense, it may also
have strengthened the collective sense of oppression and of justice.
Whether workers won or lost collective grievances, the process most
likely heightened shop floor solidarity.

## The Faces of Management

The degree of antagonism embedded in worker-boss relationships
varied depending on the level of management identified. Steelworkers, for example, were measurably less hostile toward front-line foremen as a group then they were toward upper-level supervision. Brad
Ramsbottom, like all workers, distinguished foremen from owners
and corporate executives. Shaking his head from side to side, he explained, "The company executives were seen as outside the guys in
the plant; they were a different breed all together. . . . Their interest
was not in doing the work. To the owners, workers were just check
numbers and there was no concern whether you had a job or not."

While "there was always a fence" separating the "ranks and management," Youngstown steelworkers articulated a bewildering array
of feelings toward their employers.[19] Workers knew they were "just
check numbers" and fully expendable, but expressed little universal
antagonism toward management. Managers and owners were good or

In 1956 Brad Ramsbottom (*right*) helped to post a strike notice. Reprinted from *The Vindicator*. Youngtown, Ohio. © The Vindicator Printing Company 1998.

bad depending on what they did. Jim DeChellis's ambivalence is representative of the way workers both denigrated bosses and found common ground with them. He admitted, "I don't know why we had bosses." At best, "they just made up the schedule, but they didn't know the job." However, DeChellis believed, "We both cared about the job." When all was said and done "these guys [here he included corporate executives] were just like us."

How workers felt about their employers often diverged along a fault line that separated "steel men" from "finance types." U.S. Steel (USS) worker John Varga made this distinction during our conversation.

> *R. Bruno:* Did you feel that you and other workers were part of a class struggle?
>
> *J. Varga:* I think so. With Roderick [USS chairman], they never understood the workingman. When Speer [chairman before Roderick] was there we were well taken care of. He cared about the workingman. We were lucky to have him on the top. But Speer represented steelworkers, where Roderick represented making money. With Speer gone, it was like one class against the other.
>
> *R. Bruno:* When the mills closed did you feel cheated?
>
> *J. Varga:* We were making $25 million a year, then the company shifted the work and turned a profit into a loss. Two other plants got the work. This mill [Ohio Works] was making money. Edgar Speer came out of this mill and said it would never close. But then he got sick and Roderick took over. He was out of accounting and was just a money man. He closed it down.
>
> *R. Bruno:* Why did the mills close?
>
> *J. Varga:* It was one man, Roderick. He was the man who tore us down. When Speer got sick we knew the signs were bad. I wasn't surprised when it closed. Speer actually told us the other vice presidents wanted to close it down four years ago.

Many retired YS&T and USS employees held to a similar firm belief that class antagonisms had grown more severe because steel executives in the 1970s cared less about making quality steel than did their forefathers. This belief was a fascinating mix of reality, propaganda, and confusion.

Before 1969, the area's largest steelmaker, YS&T, was fundamentally a locally operated business. The company was created in 1900 by four Youngstown investors who raised $600,000 in initial capitalization. By far the biggest employer of area labor, YS&T operated plants in Youngstown, Campbell, and Struthers. The Campbell plant alone employed over five thousand people, with another sixteen hundred at Brier Hill. For nearly seventy years, workers had considered the company a good, responsible employer with strong social ties in the community. Bob Dill offered what seemed to be a shared worker opinion about the "old" YS&T: "I thought the old owners were wise men who were working towards the future. These guys were really interested in making steel." But that all began to change in late 1969.

Looking for new investment and product areas, YS&T merged with the Lykes Corporation. Lykes was a New Orleans–based shipbuilding company that earned profits by carrying war materials for the Defense Department. They were a much smaller company with assets of $137 million, as compared to YS&T's $806 million, and the merger was described as a case of the "tail wagging the dog."[20] If size differential was not problematic enough, the management of Lykes knew nothing about making steel. Yet the company put together a proposal and found a way to convince the local ownership to accept the terms of a merger.

In order to pull off the acquisition, Lykes borrowed $150 million to buy a minority stock interest in YS&T and eventually assumed a debt liability of nearly $350 million.[21] From the beginning the signs were all bad for Youngstown steelworkers, and in less than a decade everything good turned rapidly rotten. On 19 September 1977, less then 8 years after the merger, Lykes-YS&T Corporation made the announcement that they were closing the Campbell Works.

While "Black Monday" ushered in the next decade, the attitudes of workers toward their employer had begun to change seven years earlier.[22] Under the plant's new ownership workers continued to provide a fair day's labor but gradually came to damn their employer with distrust. Lykes was perceived as a thief who raided the community treasure chest. Workers were angry at the new company management because they believed that "YS&T was a viable company" and that Lykes "didn't invest in the mill."[23] The feeling that the company abandoned steel production was widespread and was a major source of worker antagonisms.

Workers gave a number of plausible reasons for why the mills closed down, but nearly all of them blamed Lykes-YS&T for "not modernizing the plant."[24] According to the workers, the company would not spend any money on parts and repair work. Nothing new was coming in, and everything old was getting older. The most common form of maintenance quickly became "burlap and wire."[25] Workers also criticized the company for what they did with the money they were not investing back into the area. Instead of pouring the profits back into steelmaking the company was obviously "looking to find better investment opportunities and to get out of steel."[26] Bob Dill knew the company was up to something dishonest: "The company would tell me they didn't give a damn about the steel; it's the money, money, money!"[27]

The focus on generating income without reinvestment meant to workers that the company would also chase a cheaper form of labor. Speaking about Republic Steel, Tom Bergman believed, "If they could find cheaper labor somewhere else they would go." For some steelworkers, the lack of modernization and eventual deindustrialization of Youngstown was all part of a larger financial strategy by big corporations. U.S. Steel employee Frank Kerrick said the companies' plans included union avoidance: "Their idea was to break the union. At least control them better. The company actually took skilled workers to Japan and other places to show them how to make the steel better and cheaper than we could. The company used union labor to help them get away from the union."

Research conducted in the state also supported the fears that workers expressed. Lykes's cash flow earnings in the early 1970s were high, but "instead of returning this income to its steel division, Lykes made other investments."[28] Workers were particularly incensed that while they were being asked to give contract concessions to secure jobs, the company's stockholders "did real well." So well in fact, that while YS&T's earnings fell off between 1972 and 1976, Lykes was paying out stock dividends greater than the industry average. The company paid out dividends worth $8.26 a share, compared to the industry average of $5.83. They also reinvested $2.27 per ton of steel produced, while the industry average was $9.62.[29]

The workers' historical view of the YS&T Company is generally correct. The company was in better financial shape before the merger, and it did show a greater interest in modern steel production than its successor. YS&T took on a significant expansion project after World War II. They invested $1.4 million in modernizing older production units and building a new mill.[30] But the old YS&T compiled less than a decorous labor relations record. It was a fierce opponent of SWOC drives between 1936 and 1942, and sponsored a company union to break the 1937 Little Steel Strike. The group was called the Independent Society of Workers (ISW). They published a letter urging all Campbell Works men who wanted to return to work to meet at the society's headquarters in the Dollar Bank and sign a registry.[31]

During the war the company testified before a panel of the National War Labor Board against a closed shop, a dues check-off, and a one dollar a day wage increase. In 1941, the National War Labor Board filed formal complaints of unfair labor practices against the company. The company opposed the demands put forth by SWOC because "to grant a closed shop would increase industrial strife, retard the nation's efforts toward maximum production, [and] discriminate against the millions of young men in the armed forces" and to grant a wage increase would "aggravate the dangers of uncontrolled inflation." YS&T president Frank Purnell published an open letter criticizing the closed shop for leading to "inefficiency, limited output, slowed production" and for strangling "initiative, incentive for advancement and pride of achievement." He also added that "some CIO organizers [were] avowed Communists."[32]

In 1941, the company was also found guilty by the United States Supreme Court of violating the National Labor Relations Act for firing union members participating in the 1937 strike. The court refused to overturn a lower court ruling against YS&T and Republic Steel for the dismissal of thousands of union workers.[33] Agency rulings and judicial decisions, however, represented only part of the company's conflict with the federal government.

The company also fought with the executive branch of the government when it rejected the recommendations of President Truman's fact-finding Steel Industry Board concerning wages. President Purnell sent a telegram to the White House declaring that "the company cannot agree that it will be bound by the recommendations of the suggested board of third parties as to the matters being negotiated" and that if there is an industry strike a national emergency should be declared in order "that the provisions of the Labor-Management Act of 1947 become directly applicable."[34]

Ironically, the same hometown company owners who workers said were interested in local steel production had tried to merge with Bethlehem Steel in 1930. The local board of directors approved the buy-out plan by a six to three vote, which the hometown paper declared would result in Youngstown being just "another pay roll town" instead of a "leader in the steel industry of America."[35] The move, however, was blocked by a major local YS&T stockholder who opposed the competitive advantage that the merged companies would gain in the industry. The apparent white knight was Cyrus Eaton. Eaton owned 20 percent of YS&T stock and blocked the move because he was also a principal owner in Republic Steel. Eaton did not want a merged company challenging Republic's hold as the nation's second largest steelmaker. Eaton's proxy challenge generated what was widely considered to be the "most expensive legal battle in American history."[36]

But the deal was too good to drop and history soon repeated itself. In 1950, the two companies made a second attempt at a "friendly takeover," but the move was opposed by Eisenhower's Justice Department. In 1950, YS&T was the country's fifth largest steel producer. Bethlehem Steel was number two and a merger with YS&T would have raised their ingot capacity to 19.6 percent of the industry total.[37] The "benevolent" local steel master pursued a new corporate suitor until the deal was killed in 1958 by a federal court. The company was also not an unwitting victim of steel substitute products. In 1953, while still one of the nation's largest steel pipe makers, YS&T acquired an interest in a plastic pipe manufacturer.[38]

Despite some evidence to the contrary, most steelworkers still felt that the old ownership, unlike the new, was committed to the valley and at least willing to cooperate with the union. Resentment toward the YS&T hybrid was of course heightened by the mill closings and the train of events that followed. The shutdown of YS&T's Campbell Works set into a motion a series of steel industry decisions which resulted in a steady erosion of the area's industrial base. In November of 1979, USS closed its Ohio and McDonald Works and idled about thirty-five hundred people.

In 1984, Ling Temco Vought (LTV) bought Republic Steel and merged it with Jones and Laughlin, which it had purchased in 1978. The new conglomerate was almost as large as U.S. Steel and immediately began to lose money. LTV Steel lost $156.4 million in the first quarter of 1985 and soon after filed for Chapter 11 bankruptcy protection, thereby voiding its union pension and health plans. By 1990 LTV had closed all but one small bar-mill plant out of its large Youngstown facilities. As I write this, only seventy out of a postwar high of five thousand employees remain at LTV Steel.[39]

The workers had few kind memories of the people who directly supervised them. George Papalko found the foremen at Republic to be lacking "family values," and most of them were "big drinkers." As

Papalko recalled, one particular boss could be so intoxicated that workers had to act like foremen in order to get the job done. Some nights "he would come in so drunk that he would plop right down in his chair and put his head on the desk and never pick it up." On these occasions Papalko and others "would take the orders out, give the guys their work assignments, and run the whole shift."

In cases where there was a shortage of men on a crew or where foremen failed to come to work, workers doubled as temporary foremen. Mike Vasilchek worked occasionally as a "junior blower foreman" in the blast furnace at YS&T but was never compensated for his effort. He still feels outraged by this: "The big foreman would go home and turn the job over to me. This went on for years. I learned these jobs, but they only used me in a pinch."

With remarkable consistency workers dismissed the skills and intelligence of their supervisors. Discounting the importance of foremen was predicated on one central, unyielding belief: workers "knew their jobs better than anybody."[40] If foremen were unnecessary to production, then corporate executives and owners were mere investors in production. They held tremendous power and collected enormous financial riches, but "they didn't know what was going on in the plant."[41]

While workers recognized the de facto power of bosses, they would legitimize that authority only if it was accompanied by experience and knowledge of the job. Because "workers knew the job better," only "the foremen who came from production were helpful." The rest of supervision "never did the job" and consequently weren't "important at all."[42] It became apparent that workers conceived of two kinds of bosses: *us*, the boss was part of the workers' family, knew the job, and respected the men; and *them*, company men who were organized against all of *us*.

The foremen who were perceived as enemies of the workers were those self-serving ones who "were always trying to get feathers in their hats by treating the guys like slaves . . . pushing guys like dogs."[43] Jim Morris remembered that the foreman in his department was a "real son of a bitch who ruled with an iron hand and was out to make a name for himself." John Varga lacked all respect for "flunkies who tried to be bosses, but were riding on someone's coattails." Workers did not expect bosses to do physical work, but they were supposed to be on the job, actively supervising. Foremen who sat in their offices or only came around to complain about the men were

targets of hostility. Workers in the mason department at USS's Ohio
Works lampooned foremen who installed venetian blinds in their offices. To the guys in the department, this only proved that bosses
"didn't want to work" and that management used their offices as "a
place to sleep."[44]

Republic Steel employees were particularly critical of company
bosses and policies. Many viewed the company as inherently com-
bative, and worker condemnation was a natural corollary to being
mistreated. Mario Crivelli was angry about so many things that he
was unable to confine his comments to any particular abuse. We
spoke together with a group of workers in a nondescript dining room
at the Kings Inn a few miles across the Ohio-Pennsylvania border,
where Crivelli and nine other retired Republic Steel employees met
periodically for breakfast.

While many workers admitted to reading biographies, Crivelli was
one of a few workers who had ever read any labor history. He claims
to have read *Bootstraps,* the autobiography of Tom Girdler, Republic
Steel's Depression-era dictatorial owner. Listening to Crivelli convinced me that some of Girdler's infamous anti-union philosophy had
been transmitted to post-war company bosses. I have itemized the
complaints he registered in response to a question about his work:

> [1] The company never encouraged any individual worker to make in-
> telligent contributions. Foremen would just say, "We pay you to work,
> not to think." This was very common. [2] Guys would say, "Republic
> Steel sure has a lot of fresh air." This was because the wind would just
> whip through there in the winter. There was a lot of bad conditions
> there, like grease and oil that you would slide on all the time. You had
> to watch yourself all the time. [3] In the early '80s the company wanted
> you to do two different jobs. They were eliminating jobs and combin-
> ing tasks and always wanted you to do things under bad conditions
> [here he returned to a previously cited problem]. [4] The company had
> 23 vice presidents! What the hell for? [5] You know the company
> wouldn't let you do too much to help another guy. They jumped all
> over you if you got hurt helping another guy. [6] They kept you tied to
> your job. [7] The company would do anything they could to blame you,
> the worker, for any mistake, injury, or problem.

These were the kind of worker grievances that triggered shop floor
rebellion. Pure anger and frustration are not in themselves sufficient
to present the boss with a real threat of resistance. Combined with

disdain for the company, however, they are a potentially dangerous mix. Away from worker-boss confrontations, workers said what they often felt suppressed from saying in the plant: they knew a better way to do the job. To go back to Crivelli's first complaint: the company never encouraged any individual worker to make intelligent contributions. There was no greater insult or disappointment for a worker than to be treated like a pack mule. Many workers told me of their company's displays of disrespect. Shipping floor inspector Tom Kotasek barely held back his disgust toward Republic as he remembered, "the foreman held back on doing the right things in order to save on the department budget. The company wasn't interested in running the mill and making steel. I gave my foreman a couple of good ideas about making changes in the plant and the office [superintendent] just dumped mine and other workers' suggestions in the waste can! They weren't interested in what workers had to say. The guys knew their jobs, needed their jobs, and cared about what they did."

But Republic was not exceptional. The behavior of foremen at the Brier Hill Plant of YS&T was considered so hostile that Local 1462 decided to properly "dishonor" it and expose the hypocrisy of the company at the same time. In the early 1970s, the company had instituted a PRO of the Month Award to acknowledge selected workers who had made important contributions to production. Each worker chosen would receive ten shares of company stock and be eligible for PRO of The Year.[45] Now undoubtedly, many workers who received the PRO designation were loyal, strong union members. Production efficiency did not necessarily contradict worker solidarity. But giving shares of stock to a few workers each month had apparently little to do with treating all workers with respect.

Ed Mann, president of Local 1462, commented that the thing he hated the most about YS&T was "foremen who gave no value to the work a person could do and would fight over the least little thing." It was obvious to Mann that with the company "the more [a worker] gave, the more they wanted." In response to what the local leadership determined was an inauthentic attempt to recognize worker achievements, they poked a satirical but serious point at management by nominating their own managerial "ORP of The Month." Winners were honored in the plant by anonymously receiving a stale bologna sandwich and sour milk in a brown lunch bag.[46] In announcing one particular winner, the local made clear that being workmanlike was the standard used by workers to identify good bosses: "It's

been only a few short months since this upstart left the rank and file
to become top turkey and he has already shown an amazing ability to
forget how the other side lives."[47]

Workers did not automatically expect their bosses to be incompe-
tent, brutalizing, or disrespectful, and they did perceive that foremen
were sometimes stuck between a rock and a hard place. Line foremen
needed the workers to make production quotas and to ensure positive evaluations and bonuses from plant managers and department
supervisors. In order to get the job done, bosses had to countenance
worker cooperation, but at the same time they were under intense
pressure from the company to push the men harder and harder.[48]
While workers could sympathize with the plight of the boss, they
also "tried to take advantage of that a little."[49]

## Forms of Resistance

While there were worker-foremen friendships at work, positive relations with foremen were never uncomplicated. The work process
ensured that feelings of magnanimity became factors in production
and weapons of shop floor power. Getting along with the boss "was
very strategic," and friendly relations were loaded with reciprocal
obligations. As an "ingot stripper" (i.e., a crane operator) John Varga
held an essential job in the flow of work between major production
units and understood the need to cultivate a manipulative relationship with his foremen:

> I could shut the whole mill down if I didn't move those ingots. I had
> bosses coming to me all the time asking for things. I'd tell them, "You
> just have to come up and ask me," and I would do what I could. You
> see, I was in the middle and had a key job. The process depended on me
> and gave me an advantage. I once had a boss telling me that I had to get
> twenty-four ingots out of there. So I decided to push them all down to
> the ground! He came running over and asked me, "What the hell hap-
> pened?" I said, "I don't know." They understood what I was saying.
> But you couldn't just order me, you had to ask. The railroad guys had
> the same kind of leverage. If they wanted steel they had to ask me.
> They learned real quick. I only had to do it once in a while. My fore-
> man would tell bosses in other areas, "He [Varga] could put them all
> [steel ingots] on the ground in a minute and there's nothing you can do
> about it."

John's job was to use an overhead crane to transport into soaking pits the smoldering ingots made by pouring molten steel from the open hearth furnaces into molds. Once the ingots were cooled, they were taken to rolling mills, reheated, and then shaped into various steel products. If the ingots were to topple over like dominoes, they would either harden on the ground or bust apart. The loss in production to the company would be enormous. Varga, it should be noted, also thought very highly of one-time USS president Edgar Speer. In addition, he named a number of foremen whom he respected. But when his knowledge and control of the job was threatened, that admiration meant little. In those cases, his preference was to just "put them on the ground."

Workers always suspected that because of their mutual need for one another a degree of supervisory hell was ordered from above. Local unions would actually advise their members not to speak too positively "out loud" about any boss they liked, because this would put the foreman in an awkward situation with the company and of course result in a loss for the workers.[50] Much of what foremen did angered workers, but they took into consideration that "foremen were warned about being too close with the rank and file."[51] In response, workers and shop floor bosses sometimes cautiously "created [their] own alliances."[52]

One common example of how this alliance helped bosses and workers was the issue of safety enforcement. Foremen needed to prove to senior management that they were tough in enforcing the rules. So they went to workers and told them they had to "write up" a certain number of safety violations each month. The workers who were "wrote up," however, were selected by the men. They usually had excellent safety records, and the violations were always minor. Under this system, the foreman earned his "credits," no worker got punished, and the workers were owed a favor. This leveraging was explained to me by my father, who volunteered once to be "wrote up"; and conversations with other workers established that deals like this happened all the time.

Under conditions satisfactory to the needs of workers, bosses became part of the "people who work in the plant."[53] Steelworkers did not mind responding to an orchestra leader who coordinated movement, so long as when that person "want[ed] to learn something about production, [he went] to the workers."[54] Antagonisms, however, flared when managers refused to listen to workers. Management's "attitude that they were better than [workers]" drew the clearest distinction between workers and nonworkers and was a flash point for worker solidarity. When asked what united workers, many of my respondents said, "Their attitude towards the foreman."[55]

Without exception, workers complained about being treated as children who should be seen and not heard. Workers were supposed to "work with their muscles" and "bust [their] asses," but not have an idea worth voicing.[56] Jim Visingardi hated the imposed silence most of all. As he recalled, "The company [Republic Steel] always thought they knew the best way to do things, and they didn't think their workers could contribute anything. When you made suggestions they came with that 'I'm the boss' business. So you stopped telling them. This was very typical. We said, 'Damn them, so go ahead and struggle.' It wasn't a good supervision-to-labor situation. This happened over and over again."

Mario Crivelli remembered that one difficult foreman "was so bad that when he moved to one end of the department, the guys on the other end would stop working, and when he moved back the other side, guys would put their tools down." Crivelli defiantly claimed that "jobs barely got done."

Crivelli's confession points out the methodological danger of confusing cooperation with consent. It is unlikely that the observations made of workers in "public" settings ever provided a complete record of worker consciousness.[57] While workers' consciousness may have at times been out of sight, it was not necessarily out of mind. The typical worker reaction to a recalcitrant foreman was, according to my father, to "bitch like hell when he wasn't around and just ignore him when he was." Once the "bullshit" was over, workers often rejected the foreman's directions and simply "did the job the way [they] wanted to."

Going along with what they knew was a bad idea was another way for workers to sabotage the bosses. Arthur Newell told of one such experience:

Actually, the foreman needed the workers. I never knew a worker who needed a foreman. We made them look good. But when Posner [principle owner of Sharon Steel] came in he went through the guys' lockers and emptied out all spare parts. Guys would keep them there for a good reason. The millwrights would store the parts that most often broke down in their lockers so they could quickly get the parts to repair jobs and save hours of downtime. But company police just rifled through guys' lockers. The result was the millwrights would have to go into the office and request a part, then wait to fix the machine. It amounted to more downtime! But that's the way they wanted it.

Down time was lost money. Failure to stop a destructive practice could also become a direct penalty for a specific boss. Plant supervisors were fond of dictating the way a job should be done, and workers resented that they just "busted [a worker's] ass looking for trouble."[58] Sometimes, as George Papalko recalled, the best way to get a boss "off your ass" was to let him learn a lesson the hard way: "One day we were working with a live wire and the electrician said not to move the cable a particular way or you'd get the shock of your life. Well this general foreman comes down and starts to yell about getting the job done. He said he was going to show us how to fix the problem, and he reached for the wire. Well, nobody said a word about it being live, and boy, did he get a jolt—250 volts of electricity went right past his ear. It was a painful lesson, but he had to learn it. That's how we got even."

If getting even justified strategic cooperation, benign ignorance, and playing stupid, then simply wanting to survive warranted taking control of the machines. One drastic form of reappropriating the production process from "dead labor" was to shut the machine down. While few workers admitted to industrial sabotage, machine breakdowns could, as Anthony Delisio revealed, be manufactured. "If guys wanted breaks," he said, "they would bust up the equipment." Delisio worked out a deal with the production workers, "If they wanted a break I would create a breakdown for them that didn't give me any extra work." This sabotage was not an irrational individual act of anger but a planned, collective, protective practice. Workers in two units agreed to stop production in a way that covered their duplicity, increased rest time, and added no additional burden to any other workers. Of course individual acts of machine sabotage did occur, but usually to the benefit of an entire work crew.

Perhaps the most subtle subversion of the rules of production was in the creation of delays. Machine operators working in the finished-steel units were responsible for notifying an inspector (a union member) about delays in production. Inspectors were then directed to go to the office and have a delay card punched. But when the production problem happened and when the card got punched did not always occur in immediate sequence.

Inspectors earned incentives by the pieces of steel tested and depended upon a constant inflow of finished products. If material was slow to arrive at their stations, inspectors were not adverse to punching a delay card when no actual mechanical problem existed. They would declare a breakdown to avoid using up their time to earn a good bonus. But the process did not end here. When pipe or other production resumed, inspectors would often not go "on-line" until they had already tested and loaded a certain number of pieces. This way they could get credit for actual pieces tested, plus earn extended pay for official time delayed.[59] While workers took immense pride in their skills, money was the primary reason they worked in steelmaking, and it represented the value they created out of their labor. In order to earn a larger unnegotiated portion of that value, workers found ways to squeeze a dollar out of the strangest places.

Another way to "take back" what the company would not agree to give in the contract was to do personal work on company time. Workers would make arrangements with one another to put aside their assigned jobs and to invest time in personal projects. In some cases, that meant bringing equipment into the plant to be fixed or building some household item needed at home. When workers ignored company work for their own, they called it doing "government jobs."[60]

If various deceptions were not subversive enough, what should be made of the fact that every worker I interviewed admitted to taking things or knowing someone who stole from the company? Workers not only stole from the company but did so repeatedly. Some took enough material out of the plant "to build a house." This was the common metaphor, but in reality other structures were actually built. Since retirement, ex-YS&T employee Joe Flora has worked at the spring festival at St. Nicholas Church where patrons move from one steel-bordered, tarp-covered booth to another. According to Flora, "All of the steel at our St. Nicholas Festival was ripped off from the YS&T. The priest knew it, too!" Two popular items of theft were copper and brass. Some workers would use the material for fishing lures, and others would sell it for profit. But taking anything out of the plant was dangerous, and apprehension could lead to permanent dismissal.

As a shop grievance man at YS&T, Bob Dill would have had countless opportunities to represent workers before the Industrial Relations Department. He remembered the horror and senselessness of having to plead for a worker's job because some raw material used in the open hearth furnaces had been stolen. Dill told me of a man with 21 years in the mill who was fired for stealing a bucket of nickel. Nickel is important in making specialty steel. Dill lamented that, "He [referring to the worker] would take the nickel out every day, about fourteen pounds in all." Bob tried a number of moves to protect the worker, including going "to his priest to speak on the guy's behalf," and with great pride Dill assured me that he "lied like hell to get this guy's job back." But in the end the company refused to budge.

While the worker was allegedly selling the nickel for wine, some workers felt they had a right to the material in lieu of wages. Anthony Delisio thought that a primary reason for worker theft was as he said, "Guys figured the company owed you for the work you had done and that you weren't adequately compensated." Bob Dill recalled a second case that enraged him more than the first:

> Luigi would take copper out in a tow truck. The foreman told me about this and said he didn't want to fire him because the guy had 9 kids. I went over to Luigi and got on his ass real hard. I said, "You're going to get fired if you don't stop." Well, he told me I could kiss his dirty butt. He said, "I can't live on the poor wages they pay me. I have nine kids and this is my second job." I said, "You're stealing the goddamn copper!" He said it wasn't stealing, it was part of his job. He got fired and didn't get his job back. Look, everybody took something small. But not every day!

In both cases, Dill thought the workers had done something wrong, and this view was supported by the comments of other workers. They contended that they were paid a wage to work and therefore entitled to no more. But they also admitted that "everyone took something." Why? Workers admitted to theft but conditioned their confession on a "minor harm" claim. In other words, "It was stealing, but it was insignificant." The stuff that workers took they "bent over and picked it up and put it in [a] pocket."[61] It is also plausible that workers believed that their hard work did entitle them to the small pieces of property that the company would never miss.

The danger for a worker smuggling out company property was the plant gate security guard.[62] Since under certain conditions stealing could get you fired, it was necessary to surreptitiously smuggle items out in work clothes, lunch boxes, and any other detection proof container. But, as Joe Flora admitted, "The plant gate guard would barely touch your bag." The possibility of being nabbed was minimal because the "guard didn't want to find anything; you see, they were only going through the motions."

Workers did, however, have reason to fear company reprisals for stealing valuable materials that could be resold. And apparently some workers tried to market the product of their own labor. Curiously, while none of the workers I interviewed admitted to taking big items, they all knew mill hands that did. After listening to numerous secondhand accounts of brazen theft, it occurred to me that something like a mythology of great heists existed. Workers did not condone stealing valuable property, but they seemed to hold in high regard those who did.

It is therefore relevant to note that stories of plant robberies were very similar. As workers told them, they had a common plot line with familiar characters and a predictable climax. Workers may have fabricated a tale or two, but if so, they constructed a common fiction that contributed to the solidarity of the workers. Stories of thievery became working-class folktales full of normative messages.

Tales of mill theft were of two kinds. In one story, the worker has conceived an elaborate scheme to steal, but after a period of success is apprehended and subsequently fired. John McGarry told one of these:

> Stealing big things was a problem. In one case two guys would come in at night and back their dump truck up to the gate, and one of them would climb over the gate and drive this forklift carrying steel slabs, weighing a thousand pounds, over to the gate. He would then dump them over the gate. He could only take two or three at a time, or the truck hoist couldn't be lifted up. But one time a detective on the P.E. [Pennsylvania and Erie] Railroad noticed the truck and called plant protection. They stopped the truck off of YS&T property and the guys gave some excuse, so they let the truck go; but they followed it all night long. The guys went to a junkyard the next morning in New Castle [city in Pennsylvania]. But these slabs had a YS&T chalk mark branded into the slabs, with an invoice number. These guys got fired.

All the elements of a classic folktale are present: subordinate individuals involved in a clever plot to deceive and take advantage of a dominant power's weaknesses. While workers did not praise such practices, their stories reveal what all exploited groups weave into their folktales—the intellectual superiority of the subjected group. Accounts of worker theft always involved a shrewd, smart, and deceptive worker "tricking" a more powerful party. Workers were ultimately stopped and severely punished, but not before they "got even" with the company

If workers were unashamed to admit minor acts of thievery, they were gleeful to divulge incidents of supervisor dishonesty. In fact, every personal worker confession was immediately followed with "but the real thieves were the foremen." Workers contended that plant supervisors were far more villainous because they could drive their cars inside the plant. Positioning a vehicle with trunk space close to the work site presented more exciting opportunities for theft than did the bottom of a metal lunch box. While workers could heist "a couple of wrenches to use at home," the foremen covertly drove out "pipes, tractors, everything."[63] In this tale, workers are held to be morally superior to supervisors, who not only contribute nothing of value to production but also are engaged in treachery against the company. Ernest Pajatsch claimed it was a common scenario: "There was so much work being made there and charged to other departments that went home with people! And oh my God, the foremen were the worst. The day the company said it was going to start checking cars you heard more things being splashed into the [Mahoning] river! They had a ring of bosses that were making flagpoles and trucking them out at night. Top management couldn't fire these guys because it involved too many foremen." Thieving foremen were no more than crooks, who stood in sharp contrast against a working class committed to honest work, the cause of production, loyalty to the "company," and modest victories over oppression.

While workers' dedication to the company appears counter to an oppositional consciousness, on closer examination that fealty amounts to something less solid. Tales of theft ultimately (1) strengthened the workers' belief in their own superiority over supervisors, (2) provided some psychic relief from constraining work relations, and (3) indicated that workers never accepted the existing class power relationship as natural or eternal. One such tale of worker audacity that rose to the level of legend, with significant psychological implications, is presented here:

Every day this guy goes out of the plant with a wheel barrow of hay. The security guard stops him and rustles through the hay. Finding nothing, he lets the worker pass through the gate. This goes on for weeks, until finally one day the guard says to the worker in exasperation, "I don't get it. Every day you walk out with a wheelbarrow and I can't find anything in it but straw. Now I know you're stealing something from the plant, but I can't figure it out. Why the hell are you stealing straw?" And the worker answers, "I'm not stealing straw—I'm stealing wheelbarrows."[64]

Here the worker is not only duping the company man but clearly making a fool out of him.

Vulgar language and demeaning jokes were another common form of resistance. All of the workers I interviewed admitted that cursing was a second language in the plant, in union halls, and taverns—but, as my mother insisted, never at home. Only once I remember my father screaming out an obscenity. Mom's retort was a sharp punch in the arm. The brief summer I spent in the mill was an education in how to use the word "fuck" as a verb, noun, adverb, and adjective.

In most cases, foul language and humiliating jokes were directed against the foreman. The workers abided by definitive rules of verbal engagement that were class accentuated. Anthony Delisio explained, "You always cursed at management." No word more often or better described worker attitudes about their supervisors than *prick*. Of course it was never just prick. The word was usually modified by *fucking, no good, worthless,* and *little.* In the common, locker-room discourse of the mill, referring to a boss as a defective part of the male anatomy was an attempt to denigrate his humanity. But Delisio said, "If you cursed at a worker you had to be ready to fight," because it was an unwritten rule on the shop floor "to never belittle another worker." There could be cursing in conversation and cursing directed at someone else, but "you didn't curse at a worker because it showed disrespect."[65]

Along with the foremen, the work itself could be a good reason to shout obscenities. On other occasions the two would combine into an object of worker scorn. The mill's loud, unsafe, and often intense working conditions were reason enough for a strong current of foul language. But add tyrannical bosses to the mix and workers' verbal condemnation could explode into a deluge of insubordinate declarations. But unlike swearing about the job, swearing at a boss was dangerous. If done covertly, workers were protected by anonymity, but sometimes the yelling moved on stage. When a worker made the mistake of directly cursing a foreman, other workers were expected to help. Arthur Newell described how other workers could serve as "character witnesses" for the accused: "Joe Burns would curse every imaginable word at the superintendent. The superintendent would then turn towards me and say, 'You heard it, tomorrow we're going to Industrial Relations.' The next day he comes over to me to get me to testify to Industrial Relations and I told him, 'I heard nothing, and if you bring me to Industrial Relations I'm going to tell them I heard nothing. So your wasting your time.' I'm not going to turn another worker in."

Workers were expected "not to testify against each other for management" and to protect the suspect from company retribution. Protection was commonly given by simply pointing the finger of blame in all directions. According to Red Delquadri, "If there were minor problems you would try to cover them up and protect a guy. A common and clever way to cover the problem was to say some other guy, maybe on a previous turn, was responsible for the problem." If you deflected the blame correctly "you didn't mention names but you argued it wasn't you or anyone else who was responsible."

Strikes, grievances, deception, theft, and foul language allowed workers to express their antagonism toward their employers while retaining the wage-paying job on which they were so dependent. The need to resist was predicated upon dignity and a better life; the actual resisting was sometimes harmless, and sometimes a threat to profit and domination. There may have been many workers who did not understand about free-market economics, but what they had experienced firsthand convinced them that conflict was a part of the process, and in the words of one indignant mill hand, "Oh, Lord! Don't let anyone tell you there's no difference between the boss and the worker."[66]

On balance, worker resistance toward the steel companies was not dogmatic and did not preclude occasional tactical cooperation. Workers were not constantly in a "war of movement" with their employers.[67] While some form of resistance was common throughout the period studied, worker attitudes toward their employers were a synthesis of palpable anger, gratefulness, and mere resignation. Workers could be deeply thankful for their employers (something not hard to understand for white workers with little formal education, black workers shut out of other industries, and Hispanic workers who spoke little or no English); sometimes this bordered on feelings of obligation, but in varying degrees they also regularly revealed just how much they could dislike the capitalist at their side.

# 6 A Vote for a Steelworker

## Is a Vote for Yourself

Yard signs were a fixture of the local electoral season. From one end of Struthers to another, each family's personal political champion was proudly advertised on a piece of cardboard posted on a stick. Politics were personal. The names of candidates for local and county office were emblazoned on my parents' bowling team shirts. Mom particularly liked the solid black shirt with gold "Opsitnick" in script across the back. Office seekers were known quantities. They were friends of voters, and more important, workmates—at least the ones who got elected.

Prevailing postwar political analysis has linked socioeconomic identity to voting. Class has always been a marker indicating how people might vote. Typically, however, the analysts focused on party ties to class. Workers voted Democrat, and businessmen voted Republican. Working-class neighborhoods, precincts, and wards traditionally supported the Democratic Party candidate. Residents living in the wealthier suburbs were more likely to vote Republican. No political observer would dispute that the working class's loyalty to the Democratic Party was one of the pillars of national political practice. But when it came to local elections, partisanship may have been overemphasized.

Voters in working-class towns like Campbell and Struthers did consistently punch the Democratic ticket for mayor and for council seats, but not primarily because these candidates were Democrats or acted like the friends of the working class. I believe my parents and neighbors voted for local leaders according to the class identity of the office seeker or holder. It is true that from 1940 to the late 1970s, area workers consistently voted Democrat for president, congressional seats, and most statewide offices, but within the community they cast their ballot for labor union members. Instead of acting out labor's role as junior partner to the national Democratic Party, workers themselves held most elected offices in Campbell and Struthers.

Politics at the local level reveals a great deal about how working-class people tried to influence the structure and condition of their daily lives. While national and state political behavior is crucial to the fortunes of labor organizations, it is at the level of neighborhood wards that rank-and-file unionists became citizen legislators. It is also at this level where business groups, like the Struthers Businessmen Association, can be so dominant that class antagonisms are incited. This chapter examines how locally elected labor representatives in Struthers governed a traditional working-class town. Every council within the period studied was dominated by local members of the United Steelworkers of America (USWA), and consequently, for three Cold War decades, unionized steelworkers made public policy for the city. Their cumulative voting records shed some understanding on the political forms a them-versus-us attitude can take.

I have closely investigated the city council's records to look for evidence of labor-oriented governance, but what is more intriguing than actual union dominance of the town councils is the legislative efforts of labor representatives acting as agents of their class. What follows is an examination of local issues that either generated a class-based politics or failed to excite a discernible labor position. While not always overtly antagonistic to other classes, USWA councilmen characteristically acted to protect and promote working-class citizens.

As candidates for local political office, incumbents and challengers never failed to stress their working-class and union backgrounds or corresponding allegiances. "Always for labor" was the way most incumbents and challengers introduced themselves to the voters. While victorious candidates were not obligated to legislate exclusively on behalf of the working class, in order to get elected it was essential to define themselves as members of the working class.

Legislative leadership in the Youngstown area carried the union label. Unionized steelworkers were the region's dominant voting bloc, and for three Cold War decades, workers governed the city of

Struthers. They were elected to political office in Struthers and Campbell so effortlessly that local voters often considered union elections to be primaries for the citywide general contests. In both Struthers and Campbell the density of steelworkers regularly produced a decidedly pro-union elected government. From 1951 to 1978, a majority of the seven-person Struthers City Council was constituted of a majority of USWA members in twelve legislative sessions (24 years). In eight sessions, steelworkers accounted for at least 71 percent of the cumulative council seats.[1] From 1951 to 1978 incumbent turnover among nonsteelworking council members was 300 percent higher than for USWA dues payers. During this period, there were only eleven wards (N = 67) that changed over to nonsteelworking representatives in a subsequent election. On the other hand, in twenty-one wards where during this time the representatives were nonsteelworkers, ten of those wards were represented by a steelworker in the next election. Town legislators from professional or business occupations rarely served multiple terms, and only three managed to hold their seats for six years or more. Steelworkers, however, usually settled in for extended periods of governance. Tenure on the council typically ranged from twelve to sixteen years.[2]

The work-related demands on steelworkers were never absent from council chambers. How else to explain a steelworking member requesting of the citizens attending a public meeting, "Try to keep the questions down, because a man here has to go to work"?[3] The official record shows that no one dissented. It is unlikely that such a request, coming from a nonsteelworker, would have been met with such cooperation.

## Class in the Party System

Struthers's working-class councilmen were not advancing a political strategy for the abolition of the wage system. But their actions to advance the quality of life inside the community were taken with other workmates and class affinities in mind.

Despite the ability of steelworker-legislators to take a working-class stance on a particular issue, they never appeared to promote an independent labor party. Struthers's councilmen were always officially Democratic Party members, and nothing in their campaigning or policy making gives any hint of other political loyalties. Obviously steelworkers believed that a committed Democrat could legislate for the working-class. Until the early 1950s, however, there was a political group offering undiluted radical politics to the surrounding area.

In nearby Akron, the rubber-producing capital of the world, the United Labor Party operated out of five different locations. The party had chapters in both Akron and Youngstown and, according to Youngstown Sheet and Tube (YS&T) worker John Barbero, evolved out of the Independent Labor League.[4] Barbero, an active rank-and-filer for Local 1462, described the party as "a free group . . . ideas weren't discouraged." Members, including Barbero's fellow Local 1462 activist, Ed Mann, never paid dues and were held together by the fundamental socialist demand that "basic industry should be nationalized." While the party considered itself a true voice of the working class, it also "considered itself part of the American movement and wanted to get away from Marxist language."[5]

The United Labor Party never survived the rabid anti-Communism of the Eisenhower years, and none of the workers I personally interviewed knew of its existence. They were not, however, totally unfamiliar with radical politics or party organizations. The 1930s had seen a vibrant Communist Party leading the Steel Workers Organizing Committee (SWOC) organizing drive, and most workers knew that Gus Hall was the local party leader. In the 1930s, he worked in the Youngstown mills and was a Mahoning Valley union organizer. In 1937 he was actively involved in the area's Little Steel Strike. Older workers like George Papalko knew Hall and recalled that the Communist Party organized mass meetings to excite interest in the union and ran classes on Marxism. Some remembered that most gatherings were announced to the public through handbills, which among other things called on all steelworkers to "Unite! Prepare to Act! Build Your Fighting Strength! Join the Industrial Union!"[6]

Working-class appeals from the Communist Party did not stop, however, with the formation of the USWA or the beginning of the Cold War. In 1953, the party attempted to sway opinion during the Youngstown campaign for union directorship. In a lengthy printed statement titled, "The Rank and File's Stake In the Campaign for District Director" the party provided a point-by-point analysis of each candidate's positions. They offered no endorsement and con-

demned both men equally for using smear tactics. Incumbent James
Griffin was particularly criticized for "red-baiting:"

> There is one point in the Griffin program which must be condemned in
> no uncertain terms. That is the red-baiting point—"Free from Subver-
> sives." Brother Griffin knows full well that the Communists are not a
> threat either to his position in the Union, nor to the union itself. He
> knows, as many others do, that Communists have been staunch fight-
> ers for the Union from the days the first member of the Steelworkers
> Organizing Committee came to Youngstown. Why, then, does he red-
> bait us? Because he still bases himself on the class-collaboration poli-
> cies which the Steel Union has followed since the '48 split in the CIO.
> He still goes by the theory that if he gives a few concessions to the re-
> actionary forces, they will in turn give a few concessions to him. . . .
> He believes that reactionary forces will deal with him as a "respectable
> Labor Statesman," and he will be able to win gains for the workers
> without much of a fight. We solemnly warn Brother Griffin that his
> fine program will never see daylight if he continues to use such out-
> dated and reactionary tactics as red-baiting.[7]

The Communist Party's interjecting of itself into a district-level
election—a decade after Philip Murray expelled "the pinkos" at the
union's inaugural convention—suggests that radical ideas were not
unknown to workers.[8] Workers did admit to hearing plant discussions about socialism and capitalism, but most, like Bill Calabrette,
simply "didn't care about that stuff." Calabrette spoke very reluc-
tantly about his knowledge of "such things. . . . I saw literature in the
plant, but never read it," and "most guys didn't." Even if they had,
they might not have thought differently.

The official position of the Ohio Communist Party did not challenge the economic ideology of American unionism. Adhering to the
party line in Youngstown meant believing that "the most basic task
for any Trade Union—the primary reason for their existence—is to
give organizational leadership to the workers in their fight for the
largest possible portion of the wealth which they create in the Industries."[9] This justification was remarkably similar in purpose to the
reason most district locals supported Griffin: "In the every-day problems which we face, as we go about the task of protecting our gains
against the encroachments of unfriendly managements, we have

learned that Director Griffin is always there to support and guide us in whatever fight, difficulty, or project confronts us."[10]

If radical ideas appeared to be different only because they emanated from Communists, politically minded steelworkers had very little incentive to invite public ridicule over their support for an illegal organization. Given the practical need to display the right mix of "strident nationalism with diluted populism," popularly elected steelworkers probably never seriously considered campaigning as anything but Democrats.[11] For half a century in towns like Struthers and Campbell, steelworkers had always been elected under the CIO–New Deal banner. Even if the ideas of the United Labor Party had appealed to most workers, it is not likely that any of them would have run or been elected on a overtly radical agenda.

While never admitting to any socialistic leanings, Struthers's candidates for political office never failed to stress their working-class and union backgrounds or corresponding allegiances. From 1945 to 1959, Democratic candidates always included union membership in their political campaign literature. Republican office seekers also tried to capitalize on their association with local steel mills.[12] Being "always for labor" was the way most incumbents and challengers introduced themselves to the voters.[13] In Struthers, steelworkers regularly ran as Republicans. None won, but not necessarily because they were Republicans. Consider that in 1949, five Republican steelworkers campaigned unsuccessfully for council seats. In addition to representing a decidedly minority party, they also shared another defining characteristic—they were all foremen.[14] Struthers voters obviously felt a greater attachment to Democratic candidates, but they also never failed to express their distaste for would-be management politicians. Consequently, while commentators such as Richard Hamilton were correct in finding that American political parties "have never been divided along elite-mass lines," local elections could be strongly influenced by notions of class membership.[15] Contrary to the findings of voter studies of the 1950s, partisanship at the local level was not always "the anchor point of the individual's political position."[16] Struthers voters wanted mainstream New Deal Democrats for legislators and rejected their shop floor bosses as political officials.

It was an annual practice for Youngstown area candidates to appear before the Mahoning County CIO–Political Action Committee (PAC) to appeal for labor's financial support. Youngstown was one of a handful of large cities that had full-time PAC workers assigned to the area.[17] Making up a significant portion of Congressional District 19, Youngstown USWA union members accounted for 23 percent of the 1944 presidential vote in Mahoning County.[18] The CIO vote, a hefty 68 percent of all congressional votes in 1948, was particularly significant to local Democratic Party fortunes.[19] To gain such powerful backing, candidates had to demonstrate their commitment to a working-class agenda. Throughout the period I analyzed, it was customary for office seekers to be asked, as they were in 1949, whether they favored the following: (1) reorganization of the police department; (2) municipal ownership of water works; (3) low-cost housing; (4) a city payroll tax; (5) repeal of Taft-Hartley; (6) to remain free of racketeers.[20]

These commitments to class deviate from national opinion trends. Reorganizing the police department was considered crucial for labor because of the Youngstown department's history of stridently antiunion activity. On municipal ownership, public control over a vital natural resource was a favored position of the Socialist and Socialist Workers Party. After World War II, in towns like Struthers and Campbell affordable, standard housing for working-class families was minimal. In fact, as 1950 began, approximately 33 percent of both towns' housing was substandard.[21] Of course the animus of workers toward the union-busting Taft-Hartley provisions was well documented. Senator Robert Taft visited Youngstown's mills after passage of the bill that bears his name, and he was met with worker protests and vulgar derision.[22]

What was important to organized labor in the Youngstown area was not necessarily of significance to the rest of Americans. According to Gallup Poll data, the issues local working-class voters considered vital to their well-being stood in stark contrast to the opinions held by the general public. When asked in 1949 what they thought the "most important problems facing the American people" were, respondents' answers ranged from Communism at home and overseas to broad economic matters. While approximately 45 percent listed the economy as the number one problem, that category included worker strikes and labor-management relations.[23] The general public worried about workers fighting with their bosses. Youngstown workers probably did too, but not in the same way. To a steelworker, improving the chances of winning in a struggle with management (i.e., harnessing the police, revoking Taft-Hartley) took precedence over calming the public's anxieties.

In addition, national polling data confirmed that when faced with questions about class orientation, union members usually responded differently from other Americans. On issues ranging from regulation of unions to guaranteed jobs, organized workers displayed an awareness of how their objective class interests could be advanced from the choices provided.[24] But most of the general survey responses from 1948 to 1970 reveal a working-class that diverged only slightly from the profile of other Americans. Whether it was migration, artificial impregnation, epilepsy, religion, television, government administration, subsidizing of the poor, the federal government's owning and operating of essential industries, or the results of the Marshall Plan, union members appeared to think just like everybody else.[25]

Given the dominant influence that unionized steelworkers had in their communities, it is unlikely that there was much small-business resistance to local political agenda setting. As Richard Hamilton pointed out, if people live in "working-class communities, come out of working-class families, have associated with working-class children in school, and have clients who are largely working class," it should not be surprising if they support a "strong unions" issue platform.[26] Contrary to national constituency formation, Democratic Party mobilization on a local level could center around working-class themes.[27]

## Good Times, Bad Air

Not every issue was overtly treated as part of a class struggle. Perhaps the local issue that most exposed class cleavages and sensitivities was waste emissions from the valley's many industrial furnaces. The pollution claim was the city's most publicly discussed grievance against the steel companies and efforts to clean up the problem became a well-publicized reason for the mills' demise.

Since the late 1940s, the Struthers City Council had wrestled with what to do about the highly acidic "black rain" that frequently fell over the town. Pollutants emitted from mill smokestacks would mix with water in the higher atmosphere to produce a corrosive form of precipitation. Homeowners had incurred damages for many years, but the council was perplexed as to how to handle the steel companies.[28] Our home on Wilson Street, like many in the area, was covered with an imitation red-brick aluminum siding. After years of black rain the siding had grayed terribly. My father would have to hose the house down every few days to wash away the rainfall. On days when the weather was good, it was very common to walk across the driveway and feel your sneakers swishing over tiny particles of graphite, spewed across the neighborhood by mill blast and open hearth furnaces. But some days in early spring, the morning air was heavy with dew. Captured within the dampness was a porridge-like, sooty substance. You could reach up and feel the stuff floating in the air. Dad laughed in disbelief, as he explained how in the winter "you had to scrape ice and snow from your car," and the rest of the year, "you were wiping black rain off the hood."

As a child I would occasionally sleep at my grandparents' home on Powersdale Road in Youngstown. Their house sat on a hill, just a short hop from Republic Steel's Stop 5 entrance. The plant had four monstrous blast furnaces and fifteen open hearth furnaces continuously running through three daily shifts. My second-floor bedroom included a window that opened up onto a rooftop. Unknown to my grandparents and parents, I would sometimes crawl out onto the roof at night to get a better view of the "fire." Like a great holiday fireworks display, the fire was actually hot balls of orange gases shooting into the air, as one mill furnace "heat" after another was "tapped." Converting raw materials into molten steel ended a heat and released an explosion of sound and gases into the air. Each time a furnace was tapped, a hot, liquefied metal was poured into giant carrying ladles or long, narrow troughs. At night, it was the light show to beat all shows. But it was bad for your aluminum siding, car, and respiratory system.

Over the years councilmen voted for smoke abatement ordinances, made regular threats, and encouraged individual homeowners to take legal action against the companies, but never took any punitive action. On various occasions steelworking members of the Struthers City Council even overtly encouraged city residents to "submit bills for payment of damages caused by black rain to YS&T."[29] But as one city legislator explained in response to a proposed ordinance, cleaning the environment in a one-industry town is not a simple process: "[W]hat about the tax rate in this town? Who is going to take care of

the tax rate from the Sheet and Tube? Five hundred and fifty men
from the Republic Steel are out of work because of a Councilman
making a fuss about this same thing. It is an uncomfortable feeling. I
want to clean up the air in this town also. I feel very sorry for the
people in this town and I think they should sue. We are jeopardizing
our future. . . . sit back and reconsider our actions."[30]

The major concern was the economic cost that tougher air and
water pollution standards would impose on the area. Councilmen
were quick to assure the public and the steel companies that they
were "in no way, shape or form trying to push industry out." The
pollution problem was dire. But so was the prospect of lost jobs, and
lest anyone forget, "we [the town] need the Sheet and Tube as much
as they need us."[31] In the late 1960s and early 1970s, YS&T was expected to spend between $10 and $15 million on pollution control
devices. Precipitators, which drew solid particles out of the smoke,
were installed, and YS&T employee Joe Flora's strongest impressions
of mill life came from working with the new equipment: "We
cleaned up that smoke and it meant money for the valley. I worked a
precipitator and every day we would collect out of this large tank
tons of this dry, red dust. I couldn't believe what was in that smoke.
All these years, where did all that stuff go? It went everywhere. This
tank was huge. It was bigger then a house."

The expenditure was met with enthusiasm by the council, but the
community remained wary about the possible consequences of
costly cleaner air. To the residents living on either side of the Mahoning River, dark skies, black rain, and ash on the front porch
meant contradictory things. On one hand, aluminum siding was
blackened and homes discolored red had to be constantly repainted;
but on the other hand, people were working. As one local notable in
1958 commented, "[E]veryone was working and folks viewed the pollution as a sign of prosperity."[32] Good times and bad air went together.

Whether the companies were doing what was expected or enough
to clean the local environment proved to be an enduring source of political debate. Angry denunciations of YS&T's pollution discharge
record were common to council debates. Some of the most ac-
cusatory statements came from black YS&T employee and First
Ward councilman William Murphy. Murphy's district rose up from
the Mahoning River on the city's north side and was especially hard
hit by black rain.[33] His constituency included nearly all of the town's

black residences, and the area included some of the city's highest concentration of steelworking home owners. Murphy consistently delivered the council's most blistering salvos against the steel companies. At one council meeting he charged, "Sheet & Tube was pulling our leg . . . [and] doing more for other people [i.e., people in the plant in Chicago] in other areas but nothing for us."[34]

On another occasion Murphy praised YS&T for installing pollution control equipment, but in the same breath pointed out that their efforts would "not control the black rain" and to claim otherwise "would be lying to the people."[35] Murphy's arguments about furnace emissions were often neutralized by those of Tom Vasvari, another member of the council and a USWA brother. While their debate appears to contradict a class-driven political agenda, the exchanges demonstrated how vital class identity was to holding public office.

Vasvari was first elected to the council in 1959 and served off and on until 1981. His political inspiration may have been his father, who ran unsuccessfully for mayor of Struthers in 1949. Victor Vasvari campaigned that year proudly declaring that he was "always for labor" and a member of the "CIO since 1936."[36] Likewise, Tom Vasvari stressed his union affiliation and won office in 1959 with the largest citywide vote.[37] By 1966, Vasvari and Murphy had both established firm political records and took decidedly different approaches to the pollution problem. Vasvari cautioned against open criticism of YS&T.[38] In reference to the same company action that Murphy saw fit to put down, Vasvari encouraged a more conciliatory tone:

> I can understand part of what Murphy is trying to put across. . . . I feel that regardless of the wording of this resolution that Council should commend Sheet & Tube for the program that they have set upon and they indicate that they are going to proceed with this program again and I don't feel that this is going to tie Council's hands in any way or in relation to any other problems in the future. However, I do believe in a policy of friendly persuasion rather than harassment or continually giving them bad press or going in front of different news media and criticizing the Sheet & Tube Company.[39]

As debate over the company's culpability raged on, Murphy suddenly presented the city council with a citizen's petition condemning YS&T's disregard for Struthers residents. Vasvari refused to sign. In response, Murphy indirectly charged that Vasvari was not a union member in good standing and therefore was being influenced by YS&T management. The implication of class disloyalty was too serious to ignore, Vasvari needed to reaffirm his loyalties. As could be predicted, he became very defensive and proposed that the council write USWA Local 1418 to verify his union credentials. At the next council meeting Vasvari read into the record a letter from Local 1418 president Ted Katula stating that he was indeed a "dues paying member in good standing."[40] But Murphy, anticipating the move, countered with a letter from a District 26 staff representative that made the embarrassing revelation that Vasvari had worked as a "supervisor five days a week since September of 1965."[41]

Not to be checkmated, Vasvari released a second letter from a union grievance man claiming that he had only "been acting foreman" and still "participate[d] in union affairs."[42] After being challenged by other members of the council as to why he began this duel of letters, Vasvari answered that he did so "because he was accused of not being free to act since he was a foreman."[43] But Murphy did not relent and implied that Vasvari was paid by the company for lost time to attend council meetings in order to vote against resolutions critical of YS&T.[44]

The war of words between two steelworker-legislators dramatized the importance of being a working-class councilman. To act contrary to working-class interests was to be guilty of selling out to the company. Regardless of Vasvari's status or Murphy's desire to represent his north side constituency, allegations of class disloyalty were serious threats to a councilman's reputation. Vasvari was portrayed as a Trojan horse sent by the mills. He intensely repudiated that perception because if he was an agent of the company, he couldn't be acting in the interests of working-class citizens of Struthers. The cause and effect relationship was real enough, and, high vote totals notwithstanding, Vasvari knew it.

While legislating for cleaner air and water did not amount to an assault on private property, it did encroach on decisions by the area's biggest industrial property owner about how to use its facilities. More important, the resolutions and actions taken against YS&T on pollution matters were framed in a debate about best representing the local working class. The intensity of the council chamber debate over union credentials strongly implies something other than political self-interest. Bad air is bad for everyone. Neither councilman had to justify his actions at the alter of the USWA. The fact that the integrity of their actions depended on the litmus test of class reveals the value placed on class solidarity.

## The Hand That Feeds You

Perhaps the most controversial political issue facing Struthers candidates was the council's consideration of a city income tax in 1951, 1952, and 1953. The proposed tax ignited intense class sentiments and generated a prolonged political fight.[45] The legislature first endorsed an ordinance imposing an income tax on the "net profits" of Struthers businesses with offices in the city and on all "wages and salaries."[46] However, with county CIO support, steelworking members of the council proposed to amend the ordinance by placing the tax on gross business income. USWA council member and Republic Steel employee Michael Petruska felt that a tax on gross business incomes would be fairer to workers, who would be required to pay based on gross wages. As an at-large representative, Petruska was probably the town's most popular elected official. During the period I studied he held office for eight consecutive terms (1949–1965) and consistently tallied the largest number of votes. Petruska argued that the amendment was necessary because right now the "workingman [was] paying 90% of the taxes and the corporations [were] paying 10%."[47]

In the end, steelworking members of the council took varied positions on the issue but agreed with a need to tax their own incomes. Although the payroll tax was eventually passed without the amendment, workers did express a prominent political voice on a basic class-cutting issue. Thus, it is important to note that through the election of steelworkers to city legislatures, organized labor was afforded the opportunity to influence social policy directly where working-class people lived.

Another important example of class politics was a motion made by Petruska during the steel industry strike in 1959 to send letters to Ohio state representatives requesting "them to have a special session to enact a bill to give unemployment benefits to workers who are on strike more than seven weeks." The motion carried unanimously.[48]

While Petruska's record is not replete with condemnations of the steel masters or motions to socialize property, it does reveal a constant vigilance over what the companies were doing to the residents of Struthers. On many occasions he expressed citizen concerns about YS&T "dumping waste," "filling in a ravine," "dropping slag," or "damaging roads."[49] As the tax issue demonstrated, Petruska consistently backed council efforts to force the steel companies to pay more for the resources and services they used in Struthers. Along with other council members he fiercely opposed a "right to work" bill introduced in the Ohio State Assembly and supported the county CIO-PAC request to form a Fair Employment Practice Commission.[50]

No steelworking member, Petruska included, acted in any way that would threaten the corporate hold that the steel companies had over the valley. Steelworker or otherwise, an elected official knew how dependent his constituents were on steelmaking. As four-term councilman and USWA member J. L. Williams demonstrated, it could not have been easy to move critically against an industry that "pays 65% of the city's taxes." Williams felt that the council should try to make "things a little easier for the company (i.e., YS&T)," because of the city's financial dependency. Williams was the staunchest defender of the company and on other occasions spoke out about the "constant harassment of YS&T."[51] Councilmen constantly faced the problem of how to demand enlightened industrial policy from a company that was their "bread and butter."[52]

The effort of councilman Joe Vlosich is a good illustration of the pressure town leaders felt. Vlosich was a seven-time representative of the town's Third Ward and a thirty-year man with Republic Steel. He had strong words for the way YS&T hauled "that red slippery ore" around the city and rerouted its trucks so they would not "damage their roads in the mills."[53] The fact that even though the council had spent twelve thousand dollars to resurface the roads the company still persisted in operating "19 ton loads" proved "they had no concern for the streets of the city."[54]

Eventually Vlosich grew frustrated by YS&T's imperial disregard and challenged the company by introducing an ordinance to lower over-the-road weight limits. But in trying honestly to represent his predominantly steelworking neighbors, he didn't fail to acknowledge where those friends earned their livelihood. "I don't want the business of the Sheet and Tube moving out of here," declared the councilman, "and I'm not threatening them, but I am going to fight until I get this ordinance passed through."[55] This is how it usually worked. A council member could get tough with the company, but only after a disclaimer acknowledging the town's near subservient status.

Occasionally this dependence required public displays of gratitude. When YS&T's chairman J. L. Mauthe retired, the council passed a resolution expressing the city's appreciation and presented it to him at a testimonial dinner.[56] It seemed only fitting since every family in Struthers knew the Mauthe name, although they may not have known who he was. Like nearly every other child in the city, I knew the name from the sign at the entrance of the park behind Manor Avenue School, Mauthe Park. The park was the glorious home of the Struthers Little Baseball League and the best ball field in the city. The land was donated by YS&T, and in 1958 the city dedicated the J. L. Mauthe Park Baseball Field.[57] It was here where I hit the famous first home run, and here where I met ex-Cleveland Indian and Hall of Fame professional baseball pitcher Bob Feller.

No councilman was ignorant of the relationship between their constituents and the steel plant along the river. This realization exerted a drag on radical political possibilities, as best seen in the aftermath of YS&T's 1977 closing announcement. The council's reaction to the announcement, however, was foreshadowed early in 1972. At the time, there was considerable agitation about the growing importation of foreign steel. While steel imports actually declined in 1972, the volume of tonnage entering the country had increased by 81 percent since 1960, and the year before had seen a record tonnage entering the country.[58] Councilman Vlosich declared his concern about the use of foreign-produced steel by requesting that the council send a telegram to the two people he believed were most responsible for influencing international trade, President Nixon and the president of General Motors: "We all live in a steel town, and we make our lives by steel, and they are doing away with a lot of jobs."[59]

*They,* of course, were the foreign steel producers. Youngstown's overall 10 percent unemployment rate was twice that of the nation as a whole, and Vlosich's warning resonated with the public. Soon after, the council began to speak out against a domestic industry "plagued with imported steel" and unanimously approved a resolution denouncing the use of foreign steel by American companies.[60]

Older industrial towns like Struthers were particularly hard hit. Revenue from the previous year was down, and the town auditor

ominously reported, "[W]e face serious problems and should consider
all spending problems."[61] The council had been slashing budgets since
1968 and now was faced with a need for even greater cuts. The financial situation in the community grew so desperate that the mayor requested a special aid package from Washington. Under the 1972 Emergency Employment Act the city received $120,000 for re-employing
people.[62] To be sure, imported steel was not completely responsible for the fate looming over the valley. It had become painfully
obvious to everyone not only that economic expansion had ended
for the Youngstown area but that a serious retrenchment in general
living standards had begun. Foreign steel made for the perfect vil-
lain.

There were other signs of economic disaster. Struthers had been
"plagued with financial problems" since the mid-1960s, and in 1969
there were confirmed reports of YS&T's merger with the Lykes Corporation.[63] What to do about the pending merger was openly discussed among Struthers's officials, and this generated debate about
the propriety of the council's involvement in corporate affairs. In the
ensuing tussle, the council displayed remarkably contradictory levels
of class and political consciousness. The issue was raised when
Councilman Williams proposed "a resolution condemning the proposed merger with the Lykes Corporation and the Youngstown Sheet
& Tube Company." It is important to note that Williams' motion
never addressed imported steel as cause for alarm. But instead of
finding unanimity among his fellow representatives, Councilman
Murphy, of all people, strongly dissented. In explaining his reasons,
Murphy stated, "Council has nothing to do with this matter and it
has nothing to do with Council." He then offered a confusing reason
for taking a hands-off policy, warning, "[W]e [council] should not get
involved because we live here." The bottom line even for Murphy
was "[I]t is not our prerogative, this is between the stockholders and
the company."[64]

Despite having a record of company bashing, Murphy as a
steelworker-legislator believed that ownership was the exclusive do-
main of corporate executives and stockholders. Murphy, although he
was a steelworker and was elected predominantly by other steelworkers, did not act as an agent of the working class on this matter of
extreme relevance to steelworkers. But neither did he lack a sense of
class solidarity with others further down the social scale. He publicly

stated, "Sheet & Tube is blackening our homes and I have never seen such outbursts." Then sounding every bit the populist, Murphy declared that the council should "support the poor people, stop talking about the big businessman."[65]

Taken as a whole, Murphy's comments may also imply that his hands-off policy toward the buyout of YS&T was less an internalization of a nonclass ideology than a recognition of a Faustian bargain. Murphy was angry at the concern the council was showing about the change of ownership at YS&T. Why hadn't other less prominent parties been treated to such attention? YS&T had been indifferent to city residents on several matters, but now that they were in danger of being swallowed by a small, nonsteel entity, the council was worried sick. To be sure, the council had good reason to agonize, but Murphy's response implies that he was indifferent to a change of ownership because he saw it as a kind of retribution. Lykes's taking control of YS&T wasn't good for Struthers, but Murphy indicated that he thought it might be bad for YS&T's management.

As conditions in the Youngstown area deteriorated, Struthers councilmen spoke more urgently about the industrial demise happening before their eyes. Like a physician skimming over the medical report of a terminally ill patient, the council considered the various causes possible: exhaustion of raw materials, expensive overland transport, pollution control costs, old mills, Japanese steel, and so on.[66] The cumulative effect of these items seemed to be transforming the valley from a Mecca for steel producing into "death valley."[67]

Despite the tragic wrenching of a working-class way of life, no steelworker-councilman condemned capitalism. No one called for a mass movement of working-class people or encouraged their peers to occupy the plants they had given the best part of their lives to. There were no calls for a worker-owned steel mill, and the council never even expressed anger at the failure of the two-party system to save the American steel industry. Councilmen did speak of the need for a national industrial policy, a domestic contents bill, and state emergency assistance.[68] In the end, the best the council could do was to support whatever was necessary to resuscitate the American steel industry. It was hardly revolutionary. When Mayor Anthony Centofanti rose to address the city, just two days after YS&T announced its abandonment plans, he expressed the confusion and fear that nearly every steelworker's family felt:

The City of Struthers as well as every other community in the Mahoning valley is still recovering from the shock of the announcement made
September 19, 1977, by the Youngstown Sheet and Tube Company.
Since . . . this announcement was made it has been very obvious to me
that our city was going to be faced with some very serious problems
brought about by a greatly reduced financial base from which to oper-
ate. . . . How would our schools be able to maintain the quality education our children need and deserve? Most importantly, what does the
future hold for the many of you whom I represent when you find out
suddenly that the breadwinner in your home no longer has a job after 5,
10, 20, or more years of service to a company?[69]

When the mills went down permanently, Struthers's working-class
legislators were as helpless as the other steelworkers. Nevertheless,
the electoral power of union members throughout a national period
of supposedly middle-class consensus seems remarkable. If class
identity mattered less to national electoral choices, it only wavered
in Struthers after the collapse of the steel industry and the disappear-
ance of people making steel.[70]

Class continued to matter in Struthers because the community
was constructed upon the relationships of steelworkers and their
families. Undoubtedly, labor legislators were workmates of many
residents who cast affirmative votes. They were also longtime neighbors, fellow church parishioners, bowling team partners, and old high
school pals. Townspeople certainly voted their interests, but their interests were best understood by people who lived like they did. Class
was the most critical marker for determining political allegiance. A
vote for a steelworker was a vote for yourself.

# 7   Youngstown, Once Famous for Steel, Lost Its Name

By the late 1980s, the devastation of mass shutdowns had left an industrial wasteland stretching for miles along the Mahoning River. Where the view from Wilson Avenue in Campbell had been fields of steel and a sky holding the tips of burning smokestacks, now there was only cold metal and weeds. When the corporate maneuvers were concluded, the Youngstown area had lost forty thousand manufacturing jobs, four hundred satellite businesses, $414 million in personal income, from 33 to 75 percent of the school tax revenues, and some very good neighbors.[1] While most "forced retirees" remained in the Youngstown area, one morning a metal "for sale" sign appeared in the front yard where my brothers and I had once played tackle football, and my family said good-bye to Sam and Lorraine Henry, who had been our neighbors for twenty years.

All that workers had previously found bad about mill life paled in significance to the experience of massive industry abandonment. Worker resistance had been conditioned on knowledge that mill labor was the best paying job available, but after the shutdowns there was little that resistance could preserve.

Resistance to the closings did occur, and forces within the Youngstown community put up a front against the corporate boards, government indifference, and large financiers.[2] While local resistance was inspiring, its implications for workers' consciousness are ambiguous. Only a small number of workers actually got behind a community buy-out idea because social and economic factors mitigated the direct harm done to senior employees and because most workers believed that the efforts to reopen the mills were futile. During the shutdown phase (1977–85), for example, many workers were eligible for retirement with 100 percent of pension. It was common for workers like Chuck Standwood to say that while they were personally in good shape, they were angry and concerned about the younger men. Except for a few jokes, Standwood said very little during our time together. But he explained, "A lot of us didn't mind the shutdowns because we could take the buyout and we had our years in. It was the guys with less then thirty years that got hurt."

While "times were good," Viola Modarelli pointed out, most workers were not aware of a "struggle of classes," but "when the mills shut down then people began to see things differently." But as Merlin Luce observed, for most senior workers, increased class consciousness did not translate into militant collective action:

> They should have had a sense of solidarity. Sometimes they did, sometimes they didn't. The most important thing that has happened in this century to the people of Youngstown was the destruction of the mills that put thousands of people into the street. Every family in the valley was touched in some way by this. The most important thing that happened to all these people was decided by a small number of rich, white men of another class. You don't do that unless you have the power to control everything. It comes from property ownership. Why couldn't the workers vote on the closings. The doors got shut and that's it. The democratic system allows them a say in their political destiny. Why couldn't workers have a say in this decision?

I asked him, "Are you saying that workers were most class conscious when the mills closed down?" Luce replied, "Yes, they really knew where they stood then." The first workers to leave the plants for good indicated where they stood by tossing their hard hats and safety boots into the Mahoning River.[3]

Ironically, the climactic battle of the closing period was waged over a worker-community attempt to buy the very property that workers had always denied having any ownership claims to. The idea of buying a closed plant was first proposed by workers of Youngstown Sheet and Tube's (YS&T) Brier Hill Local 1462 in the aftermath of the Campbell Works announcements.[4] A month later a conference was held at the First Presbyterian Church to discuss the idea of employee-community ownership, and soon after that the Save Our Valley campaign was initiated to raise funds from local citizens and to pressure the federal government to provide matching grants. Within a year of the campaign's start, 4,138 Save Our Valley Accounts had been opened up at nineteen banks, with $4 million already committed to the cause.[5] Even so, while most of the workers I interviewed thought the idea of a buyout was worth a shot, only one opened an account, and none of them ever participated in a single collective act to reopen the mills.[6]

There were, of course, workers who did act. On the morning of 28 January 1983 thousands of workers rallied at the Local 1330 union hall to hear speakers urge U.S. Steel (USS) to sell its Ohio Works to the community. USS headquarters were a short walk from the union hall. After a long list of politicians spoke, Local 1462 president Ed Mann took the podium. He asked the racially mixed audience of workers if they were "going to make an action" or were "going to sit and talk and be talked to?" Mann made clear that "the action is today," saying "[S]teelworkers in Youngstown got guts and we want to fight for our jobs." He ended his remarks with a personal commitment and a call to action: "Now I'm going down that hill and I'm going into that building. And any one that doesn't want to come along doesn't have to but I'm sure there are those who'll want to." Workers quickly left the union hall, charged the company's offices, and smashed open the front door. The occupation lasted until early evening. [7]

Mobilization around the buyout strategy was limited. Many older workers were afraid an employee-led purchase would threaten their pensions; terminated workers were concerned that they would forfeit or be denied government assistance; and the economic security of home ownership and unemployment compensation softened the blow of plant closings. Another crucial consideration was the likelihood of government support and corporate cooperation to assist the buyout plan. John Varga, a member of Local 1330, said that "the guys were hot for" a worker buyout but that USS refused to sell the Ohio Works. In the opinion of many workers, steel companies "wouldn't give the working man a chance." Varga went on to do the math: "You know, one year after we closed, they tore it down [Ohio Works]. You know why? To keep the workers from reopening the plant. Because the company knew that if the guys bought it they would make steel and earn a good profit. We could have made $25 million a year on just four furnaces working two turns with 180 men. But they just wanted to shut it down."

Also the International Union showed little enthusiasm for the buyout, and this prevented the average rank-and-filer from getting behind the idea. Employee ownership had few adherents within the union hierarchy. The record of employee-owned companies was not very encouraging, and unfortunately, local efforts to try to construct a new model of union representation clashed with the International Union's respect for the service approach that had always paid big dividends. On a number of occasions, International and District 26 staff practices ran counter to the regulatory, financial, and legal strategies of Local Unions 1462 and 1330 to reopen the plants. While a feasibility study was being done to determine the possibility of reopening the Campbell Works, the International issued a public paper casting serious doubt on the ability to "invest enough money to buy the closed facilities" and the likelihood that "the Federal government [would] grant, or guarantee loans, for most of the money needed to modernize the Works."[8]

The institutional forms of labor resistance deterred many workers from rallying behind street-level confrontations. In the past, elected leaders would negotiate contracts, and except for cases of wildcatting, grievance procedures were used to right any wrongs. Official strikes happened a lot, but since 1942 they legally occurred after a contract expired, and besides, most workers were never needed on the picket line.

It seems that in the abstract, workers endorsed the campaign to reopen the mills, but as Chris Cullen of Local 1331 explains below, most did not think it was realistic:

> *R. Bruno:* Should the workers be able to stop companies from doing what they did in Youngstown?
> *C. Cullen:* Sure they should, but the worker doesn't have that kind of power.
> *R. Bruno:* Should the workers have control over the introduction of new technologies?
> *C. Cullen:* I think they should be consulted because they know the work.
> *R. Bruno:* Should workers have control over investment decisions?
> *C. Cullen:* I don't know about control, but they should be listened to.
> *R. Bruno:* Well, should workers be given an opportunity to operate their own plant to keep it from closing?
> *C. Cullen:* They should definitely be given a chance to buy it.

> *R. Bruno:* Did you take out a Save Our Valley Account?
> *C. Cullen:* No. Never thought it would work.[9]

Cullen was no friend of the boss, not now anyway. Both he and his wife, Katherine, had become active unionists. His forced retirement gave Cullen a new found urgency to oppose the boss. Katherine said she had always understood the need for the union, but after her husband's pension benefits were temporarily withdrawn (due to LTV's bankruptcy), she found herself becoming surprisingly militant. "Workers had paid their dues," she exclaimed, "[and] now the company was trying to take away our future." Chris and Katherine Cullen's pronounced antagonism toward the company had enough influence on their two daughters to turn them into active union members. One was even a union executive board member.

As our conversation ended, Cullen displayed a poem that one of his daughters had written on the occasion of local militant labor leader Ed Mann's death. The poem was not about Mann, but its prose suggests the symbiosis that existed between a working-class community and its individual workers:

> Youngstown was a famous place, making steel at a steady pace
> Smoke would billow from smoke stacks high
> Floating and drifting across the sky.
> The men would work night and day,
> "This is hard work," you would hear them say.
> The mills broke down in need of repair,
> Little money was spent, even though it was there.
> Imports took over. Oh, what a shame.
> Youngstown, once famous for steel, lost its name.

The Cullens, like other area lifetime steelworkers had found an emotional identity more appreciated in retirement than in the days when "good money" was to be made.

But despite Cullen's belief in the need for collective worker action, the practical barriers to political mobilization that he and his peers faced certainly contributed to the rather tepid response from workers. To be sure, the early days of the Campbell Works' closing did generate a significant worker response and a great deal of "fairly militant action." As local labor attorney and activist Staughton Lynd observed, however, as the struggle wore on "collective outrage dim[med] and personal survival [took] over." Without a quick resolution, resignation set in and the "rhetoric of struggle [was] replaced by a rhetoric of benefits."[10]

Buyouts and political mobilization were a hard sell for average rank-and-filers, particularly when their International Union leaders were sending negative signals. Workers had paid a heavy price for the union cause and acting contrary to the advice of institutional leadership was tantamount to treason. While most workers harbored deep grievances against their employers and were inclined to support greater worker control over the business, they were disinclined to get involved in an employee-community ownership drive.

## Fair Shares

Perhaps the controlling reason for a lack of rank-and-file militancy was an acceptance of the principle that the mill belonged to the legal owners. With two notable exceptions, the attitude of most area union workers can be broadly defined as a willingness to work hard and leave mill ownership to others, if they were ensured a "fair wage." According to many of the steelworkers I interviewed, free enterprise was built on production for profit, and that in itself was tolerable if "everyone got what they earned." Except for a group of union activists in Locals 1462 and 1330, most steelworkers found no great fault with the forty-year-old institutional structure of labor relations. Brier Hill Local 1462 had been a hotbed of support for the national rank-and-file caucus in the steelworkers' union, the Rank and File Team (RAFT). RAFT had challenged the policy-making record of the International Union leadership during the major downsizing of steel operations in the 1970s. Local 1330 was generally recognized as historically having the most progressive leadership and policies of all the Youngstown steel unions.[11]

According to many workers, the existence of the mill had nothing much to do with them. The mills represented the capital investments of wealthy men, and most workers had no idea where these people got their money. Someone else made the mill possible, so someone else had earned the right to own it. But when asked whether the owners had a right to shut down the mills, most workers responded that they had a legal right, but then added that "it was the wrong thing to do." I believe workers offered this moral judgment because they believed they had a right to keep a job they had worked at for most of their adult life. The owners, however, had no such moral claim to those jobs.

It was a simple ethic. Workers who did a "fair day's labor," according to Al Campbell, deserved a share of the profits: "People talked as if the union was selfish and foolish, but all we were doing was getting some of the profits. We didn't want to break them. We just wanted some of the profits." Steelworkers were aware of the potential economic destruction of shutdowns and had no illusions about the company's disregard for their sweat and sacrifice. But the enormity of the companies' decisions did not overrule the rights of property ownership.

Steelworkers also knew that the legal and economic system clearly recognized the rights of two parties to an employment contract: the workers and the owners. Contracts had incrementally grown better. The relative success since 1946 of collective bargaining agreements seemed to confirm a secure place for a working class represented by a union and forestalled the need for more radical alternatives. Signing a union contract seemed the most efficient way to raise the quality of life for thousands of working-class families. By 1959, steelworkers could boast of higher hourly earnings than their counterparts in rubber, oil, mine, and auto production.[12] Workers knew where their power resided and generally accepted the legal and social bargain of industrial work. They also made the same distinction that Tony Nocera did between ownership and workers' compensation: "Union property was ours, but not the plant. We had nothing to do with that. We had rights under the contract."

It is plausible that belief in the legitimacy of this social contract limited workers' ability to take the perceptual leap to employee ownership. After all, the steelworkers were mostly high-school-educated people with generational ties to the mills who had reason to believe that the system would provide a good living. Most Youngstown steelworkers expected to work in the mills, and even in the shadow of economic desolation few had unfulfilled expectations.

Nothing, of course, had come easy, but the struggle paradoxically only reinforced a working-class faith in the industrial relations system. The union had impressively staged six national strikes from 1946 to 1962 and could lay claim to the largest postwar walkouts.[13] Workers agreed that they could and should have gotten more from their working years, but only a handful were unsatisfied with what they attained. A major exception to this satisfaction quotient was the resentment of some black workers. While many white workers expressed some positive personal attachment to the mill and their jobs,

black workers found that more difficult. I asked Kenneth Andrews, a
black bricklayer helper, "Did you ever view the mill in any sense as
your property?" "No," he said. "I couldn't possibly feel that way.
They [the company and the union] wouldn't allow me. Even if I made
better money I would feel like a token. I couldn't feel a part of nothing. How could you do it in an all-white department?"

If three decades or more of wage labor made a worker more secure
in retirement, it also made his work the source of a lot of other peoples' wealth. As Tony Modarelli reveals, workers were capable of
clearly stating the nature of their exploitation: "The company would
give you so much in pay and, let's say, a five dollar bonus, while
you're probably actually making fifty thousand dollars for the company and they put the rest in their pocket. Your five dollars produced
fifty thousand in wealth for the company. The workers generated all
that wealth. If the guys don't work the company can't make any
money. We had a saying, 'As long as the millwrights were sleeping
the company was making money.'" Russ Baxter was never confused
about who was responsible for the wealth created in production: "My
attitude was always that the wealth created here was the result of the
workers. I told many a boss that if it wasn't for the workers they
wouldn't have anything to do."

Things of value were created by the "working man," and "without
the workers you'd have nothing," so consequently "workers should
get the most money."[14] The mill closings seemed to remind workers
of everything good they had put into their trade. They took great
pride in the "monuments of steel" produced by their labor, and when
"outsiders" tore it all down they had little generosity for the contributions of management. For George Porrazzo, it was not just that
"workers made the profits" but that the supervisors were "nothing
but parasites . . . living off guys like him, and if it wasn't for the
workers [they] couldn't get paid." But the actions of parasites were,
nonetheless, also part of the collective bargaining structure, and
workers could do little more than "fight for the best possible contracts without a strike." Porrazzo said, "No one wanted a strike. We
can only negotiate, we can't set policy. Article 15 of the contract said
the company had the right to manage their business. And we weren't
major stockholders. We can't tell the company if the mill is prof-
itable you can't shutdown, and we can't set the price of the steel or
anything like that."

Porrazzo may have wanted a greater say in the company's business, but like most other workers, he knew, "The company has to make a profit for us to have something to bargain over." Workers did not translate exploitation into a rational need for a radical change in ownership. As Tony Modarelli made clear, even if workers believed the steel belonged to them, they just wanted more of what they earned. I asked him, "Did you feel that the steel made was actually your property?" Modarelli replied, "You know, that was the reason for the wage and benefit demands by the union. We wanted the company to share the wealth, to spread it around so we could have a better life."[15] Even after wholesale economic destruction workers, ultimately, just wanted their "fair share."[16]

Some workers had never accepted the divide between ownership and production. While personal beliefs mattered less than performance, there were workers who identified their work relations as systematically hostile and thus promoted a more radical change. Pockets of a more extreme class resistance existed openly within Locals 1462 and 1330, and these union "dissidents" agreed with Ed Mann's belief that

> industrial peace will never come . . . as long as the present system of industrial production obtains. Human nature will not change. Capital will continue to want all it can get, and labor will continue to want all it can get. And on both sides they will fight to get it. No, the lion and lamb will never lie down together in vegetarian pastures. . . . Either capital will own labor absolutely and there will be no more strikes, or labor will come to own capital and there will be no more strikes. Personally, I think labor will come to own capital.[17]

While most workers did not equate an end to capitalism with the elimination of their exploitation, the shutdowns did radicalize the views of many. John Pallay and George Papalko exemplify the transition in consciousness that the shutdowns provoked. Pallay said, "Looking back on it, I felt cheated. I should have made more money and shouldn't have had to always struggle. I put my whole life into the mill, and I think now it really was the workers' property." Papalko added, "I thought it [company ownership of the steel] was none of my business while I was working. I had a steady job, good pay, and it didn't matter what they did with the steel. But now in retirement my feelings are more hostile."[18]

The grounds upon which that antagonism was born constructed a framework for class struggle that would, after Black Monday, leave workers with little justification for militant resistance. The shop floor and industrial conditions, circa 1977, were dramatically different from what had existed in 1937. In 1977, workers were no longer picked for daily labor, foremen were thoroughly discredited gods, scabs were not replacing union workers, and when the plants shut down, workers were not physically assaulted. Most important, there were no jobs to hold and consequently, by contract, the companies could do what they pleased with their property.

Even the most militant union activists never believed that an employee-operated mill could stave off significant job loss. Merlin Luce was a staunch proponent of worker collective action. As he recalls, fighting back had more to do with exacting a price for job loss: "Worldwide demand for the product [steel] had reached a limit. The mill owners and financiers saw reduced returns on their money and they knew that the American worker would not fight back and try to stop them. More money could be made in other areas and there was no union solidarity to make the process more painful to the capitalist." As one of the heroic leaders of the workers' fight against the shutdowns, Ed Mann also had no illusions about reopening the mills: "Corporations couldn't care less about the people. They were not our saviors. I didn't believe the mills could have been opened, but a takeover by workers should have occurred. Not to save the plant, but to send a message that a social responsibility existed."[19]

As a result of the above, Youngstown steelworkers were shocked, angry, and ready to consider a different form of industrial relations, but they had no framework for doing more than expressing their emotions and appealing to government sources for help. For some workers, such as union officer Brad Ramsbottom, a change in consciousness just came too late to matter: "I think I should have been more militant in my union activity. Some guys would say to me that we were too cozy with the company. I wasn't a great union man. I should have been more radical."

In retrospect, a potentially more radical rank-and-file consciousness was short circuited four years earlier by the actions of then International Steelworkers president I. W. Abel. In 1973, Abel negotiated away the union's right to strike for a guarantee from the steel companies that they would not lock workers out and in the case of a bargaining impasse would submit all issues to binding arbitration.

What was known as the Experimental Negotiating Agreement (ENA) governed all industry-labor talks from 1973 to 1982, including the period in which the Save Our Valley Campaign was raging.

While the economic consequences of this decision are debatable, the psychological effects of a major industrial union formally giving away the right to withhold their collective labor were chilling for class militancy.[20] In reaction to the ENA, Youngstown workers collected seventeen hundred signatures on a petition opposing the International's decision.[21] While steelworkers never wanted to strike and many never even graced a picket line, they were adamant about maintaining the right as a weapon. The union was born in the ashes of a heroic strike, and everything from pensions to work rules was owed to labor's voting with their feet. Augustine Sanchez was one worker who thought that the union made a mistake by agreeing to the ENA: "The first thing they did was to give away our right to strike. We should never have done that. In the past if we had a problem we would walk out . . . but now we can't strike and the company doesn't fear us. It was losing the right to strike that weakened [the union]. That was a threat we needed. The company did what they wanted after that."

Workers relied on the strike, because coercive means were the only ones that could compel the companies to treat workers with dignity. If they had "to go on strike for ten cents," and if "every two years [workers] had to go out" just to avoid having the company "take things back," a labor boycott was an essential class weapon.[22] For many workers, then, the single most damaging thing to class solidarity was "when they [the International] put that no-strike clause in the contract."[23] After forsaking the most effective, supported, and militant collective action available to the workers, it was perhaps unrealistic to expect such an affirmative show of force to automatically reappear.

At times workers fought with great intensity against a group of people they only marginally distinguished from themselves. Workers knew that company bosses, executives, and stockholders lived better lives then they did, and that "big money" men didn't deserve all those advantages.[24] In the same year that YS&T dropped the bombshell that they were closing the Campbell Works, six senior company executives were earning over $100,000 in salary alone.[25] But for a few exceptional cases, workers conceptualized nonworking bosses and owners as either luckier or just more fortunate people. George Porrazzo explained the differences between workers and company management: "Where was he born? What family was he born into? I

could never be the president of any company because my parents
were born in Italy and I was Italian."[26]

Ownership and plant management were more a product of lineage,
education, and luck than they were a logical part of the structure of
capitalist production.[27] If class differences did not automatically em-
anate from a rigged economic system, that did not mean that labor
was treated fairly. Workers knew the companies had little respect for
what they did and particularly for what they thought. Yet for the majority of workers, the differences embodied in their employment relations did not always rise to the level of class ideology.

With numerous contradictory exceptions, workers seemed to have
constructed a functional contract model for explaining their work relations: workers, collectively present in the form of a union, and
owners and managers, in the form of a company, agree to make steel.
In the words of Republic Steel employee Tony Nocera, "workers
worked for the union and the bosses worked for the company." Both
parties had a legitimate role to play in production, and according to
Jim Visingardi, responsibility for the development and success of the
steel industry had "to start with the Wicks and such who started the
mills here. But it wasn't until the unions after the war began to have
influence that we began to get a fair shake. The demand for steel was
there. You have to give big, big credit to the management for investing here, and then to the union for what they did. I think the credit
has to be shared."

For most steelworkers the objective differences between them and
management were essentially functional. While workers made the
steel, supervisors went "out there and [got] the customers" because
they knew "how to sell the steel and where it [would] go."[28] No mill
hand doubted his importance to the success of the industry, but very
few disagreed with the notion that "without the company [they]
would have no job."[29]

To some older workers who remembered the terrible pain of the
Depression, the existence of a job was such a blessing that it seemed
like a gift from the company. John Costello was one such worker
who found salvation inside the massive metal walls of Republic
Steel. In their small kitchen, Costello and his wife shared stories of
the early days, and the "old guys." The people in those stories began,
as many workers did, with nothing. Pointing his long index finger in
my direction, Costello explained, "I started in 1936, but what misery
that was. I went back and forth to the mill for six months without

getting a job. So finally I told my mother that today I'm either going to get a job or get arrested. In order to get in the plant I had to sneak past the guard, and I went into the employment office." He got a job and never gave it up.

Once desperate for work and nearly exhausted from trying to get hired, Costello felt that the mills were "company property and they [the company] were good enough to give [him] a job." But Local 2163 president John McGarry used a metaphor of boxing to explain his view of the relationship: "Management had a job to do and the union had a job to do. Those jobs are entirely different, and the rules you go by are entirely different. Management tries to show a profit out of production and the union tries to bring a good day's work and to protect their own. It's just like a fight!"[30]

## The Good Life

Although for many Youngstown steelworkers that fight went on for three or more decades, they could not be considered economically or politically radical. Yet they developed relationships, defined by their class status, that had enduring influence over their identity. Many workers were unconscious of that influence. I was struck by how class oriented the interview responses were, although none of the workers used the term class to describe what they had experienced. After enough home-baked cookies and dinner invitations, it became clear that a steelworker's class identity was taken for granted. It was not something workers grappled with intellectually. They simply knew it.

Despite prevalent academic theory about the decay of working-class culture in America after World War II, what workers remembered about their daily practices revealed an exception to social homogenization.[31] Inside the boundaries of the community workers took care of one another. With little geography to separate them, each worker was visible to the others, and exposure inevitably brought intimate contact. At all hours of the day workers went undramatically from home to plant and from plant to home. Each time they persistently moved back and forth over familiar ground, workers took with them a bit of family, community, and work. A working-class life was constituted by rotating work shifts, mandatory overtime, grease-stained clothes and hands, freezing winds in open pipe yards, broken locker-room showers, union picnics, neighborhood churches and taverns, social halls, front porch chats and backyard visits, tear gas in the face, trips to the unemployment office, layoff notices, and plant closings.

Over years of slow improvement, these elements merged into a common definition of what it meant to be working class. The way that my parents and other working families lived was not just a survival technique or an economic fact of life; it was instead a genuine form of the "good life." Working-class relations within a capitalist system juxtapose two conceptions of a human society, and according to David Stratman the contradiction "is between the values and vision of human life of the people who create human society and the values of the people who exploit it."[32]

In Youngstown, the working class at their best moments practiced a form of human interaction conducive to building a more equitable and just society. They valued cooperation, mutual aid, collective work, common needs, personal dignity, and equality of condition. Neighbors were expected to be cognizant of each other and provide mutual support, usually without being asked. A selfish act of individualism was the inglorious badge of a "scab," and human freedom was defined less by willed decisions than it was by acting in ways that preserved the past for the next generation. What steelworkers expressed through their relationships was nothing less than a nonexploitive way to live.

What ideal version of working-class society ensues from the relations of class? I believe it is a society not of strangers transformed into citizens, but of people known as community members. It is not primarily about individual rights, but about collective destinies, not in the name of narrow self-interest, but on behalf of what human beings need to live with dignity. It is a society of intense social relationships developing out of the exigencies of an ordinary life.

It is also a society that prioritizes where people work. The benefits and blessings of modern society are realized only when workers take a measure of power from their bosses. A quality working-class way of life then, unfolds on the plant floor, in the union hall, and throughout the neighborhood. Every worker is a community member and not merely a representative of a particular set of interests. Neighbors are linked by class affinities and embedded in social obligations that go beyond loyalties to abstractions such as law and order.

The Youngstown area was a better place because steelworkers built and protected their communities. When those same communities lost the occupational bases for bringing people together, much more than property taxes were at risk. It seemed to me that in the dawn of a post-steelmaking Youngstown, only one vision of the future survived; corporate profit by any means necessary. In 1988, shortly after hearing about my father's hospital visit, I wrote of my fears for a nation that lacked a working-class vision of the possible:

> When U.S. Steel and Youngstown Sheet & Tube facilities shut down in 1977, the ripples extended far beyond a few steelworkers who lost jobs. Storefronts were darkened, school doors chained and windows boarded up. Good neighbors moved away. They may have found better places to live, but while anywhere in America is still America, not everywhere is home. If mutuality of effort and community bonding are still necessary to create a trusting electorate, then our cherished Republic has much to fear from a chained factory gate and the "for sale" signs that follow in front yards.[33]

Perhaps the authenticity of the working class is best substantiated by the fact that Youngstown area steelworkers never completely stopped living like working-class members. Genuine class ways of life did not collapse into a mass consumer culture. Despite social science contentions of a classless society, Youngstown steelworkers were the nurturers of a post-Depression, twentieth-century working class. While it was true that communities sprouted some "for sale" signs and opportunities for interaction were reduced, retired workers continued to embody their previous class relations. In America's postindustrial age, many of Youngstown's steelworkers still live close to one another and socialize together on a regular basis. Friendships are not just relived in memories but reinvigorated on golf courses, vacations, family occasions, and at local union senior citizen affairs.

Many workers have died and others were lost to the community. Cora Sanchez, the wife of a retired YS&T employee, has kept track of deaths among the steelworkers. She spoke of a neighbor who committed suicide after the company he had worked for took away his medical benefits and four hundred dollars monthly supplemental pay. The poor man would walk around the house saying, "What happens if the toaster breaks? What happens if this or that happens? We don't have any money." Distraught by the thought of being penniless, he shot himself. When Cora Sanchez and I spoke in the summer of 1993 she had been searching the daily obituaries from the beginning of the year. In a little more than six months the total fallen reached 232. Hopefully, this book will keep alive the names on her pad.

Except for these sizable losses and the abandonment of the mills, life in the steel communities hasn't changed much. What was fundamental to the class identity of these workers endures. The intimate relations of class, which nourished and empowered employed steelworkers, have never been replaced in retirement by a different set of human relations.

## Memory and Class

While most workers struggled to understand the reasons for the mill closings, none of them doubted that they had been wronged. Some even became more class conscious and many now understand the past better.

As a result of our conversations, some of the people I interviewed, such as my father, appeared to discover how class influenced their lives. I hadn't expected it, but the process itself seemed to open up the class dimensions of their life. When I first approached workers about talking with me they universally expressed surprise that anyone would think their lives were interesting enough to write about. My father was typical of workers who when asked about their lives answered unassumingly, "What can I say?" After all, he had not governed a nation or started a business, or played professional sports. Why then, would anyone find important the daily contours of his working life?

But with some explanation and cajoling on my part, every worker I spoke with quickly opened up and clearly relished perhaps the only opportunity they would have to be the subject of history. It was certainly, for most, the first time. Walter Donnely was one of the few workers I interviewed who spent some time as a foreman as well as a production worker. Donnely spent most of his mill years in the mason department as a bricklayer helper, bricklayer, and as he noted, because the company needed a "black face," finally a bricklayer foreman. He spoke freely with a deep baritone voice about his hatred for the mill. Donnely claimed that the day he left YS&T, he "never looked back." His expressed disdain for the mill came as no surprise to me, but what he said next did.

Donnely told me, "Before today I had never talked or even thought about my mill years." He had working-class friends and experiences, but they were as mundane to his life as breathing or putting his mill clothes on. Donnely was telling me what so many other workers would, that their working-class lives were rarely examined by them or anyone else, including family. Donnely was no different from any other worker in lamenting that I was the first person to ever show an interest in his working life. Not even his children, who had gone to college on the wages of a union worker, showed any interest. Letters, pictures, journals, check stubs, and assorted items from workers' mill lives lay untouched in the bottom of abandoned boxes or, worse yet, were discarded from lack of interest. So many of these workers had nowhere to go with their stories and, consequently, had never really appreciated how class had shaped their lives.

I was one of those children of working-class parents; I had gone off to college and never asked questions about how my family's life was possible. But now I was asking, and something marvelous was happening. One day after I completed a number of long interviews, my mother took me aside and told me how my father had spent the day. Her account amazed us both. Dad "went down the street" to talk to another worker on my behalf. Mom claimed that she could never remember him "ever doing anything like that before."

But this quickly became routine. After every interview session, my father and I engaged in long discussions about an assortment of work topics. His show of interest in my research was unexpected and, for me, tremendously satisfying. I was so impressed that I began to describe these conversations in a journal:

> Yesterday I had two animated conversations with my father. The key thing is that neither of these talks were solicited by me. He initiated both of them after hearing me speak of the things I had heard during the interviews. Usually Dad needs a lot of promoting to carry a conversation, but not here. He went off without a push. His body sat erect in the chair, his shoulders flew back, eyes widened, arms worked liked a chalkboard pointer and Dad spoke with authority. I was surprised.

Each day brought another talk, and with every session he told me more about being working class. He was no longer doubtful that he

Like so many other workers, my father
doubted that he had anything interesting to
say. But as this project developed he discovered that he had a story to tell and a son willing to listen.

had anything important to say. Soon "Dad was waiting for me to
come home [back from interviews] to talk more about this stuff." On
one occasion, I interviewed four other workers with Dad present in a
busy strip-mall restaurant. At first my father said very little, but by
the end of the session he "seemed to have reached an apex of pride
and understanding about the union and couldn't hold the thought
in." Suddenly, he became the interviewer and began to ask questions.
We went on for hours through the clank of dishes, until the waitress
had enough and threw us out.

Dad's assumption of the role of investigator was spontaneous and
completely out of character. He was not only identifying many
buried elements of his class background, but he had begun to frame
his own questions about the past. It seemed to me that a transforma-

tion had occurred, and I made a journal entry about that: "I kind of felt like we had traded places."

Workers always knew they were working class and that the injustices done to them were because they were working class. But after the mills had closed and until someone had asked them about their experiences, many workers would have found it difficult to speak of class relations. Now, at least, my father was finding out old truths about his working life. My journal entry read:

> Came home today from a session and found my father's tape recorder sitting on the dining-room table. What I found inside made me want to cry. Instead, I just smiled, listened, and felt very good. The tape inside was "Tape 3—Job Done, Physical Set-Up, Working Shift." One of the early tapes I did with Dad about working in the mill. I never realized he ever cared to listen.

# Appendix

## Discussion Questions

1. Did you know any "bad" union members? What did they do?
   Who was the best union man?
   What made him the best?
   Did you go to union meetings?
   Did you walk a picket a line?
2. How did workers help each other at work and away from work?
3. Where did you go to spend time with other workers and their families?
4. When you were with other workers outside of the shop, what did you talk about? What did you do?
   When you spoke of the company/the union what was your attitude?
5. What experiences helped you to realize your dependence on one another (i.e., other workers)?
6. How did the company try to weaken the solidarity of the workers? In what ways was the company hostile/friendly to the workers?
7. What was the worst thing the company did to the workers?
8. What did your neighbors do to help you during hard times (i.e., the 1959 strike, the shutdowns in 1977)? What did you do to help other workers?
9. Who were your closest friends? Were they friends from the mills? Did you have any close friends who were not steelworkers?
10. Did you join any social or fraternal organizations? Were you active in local politics?

11. When was the union the most active? Did you go to union meetings? Why? What union or collective activities did you participate in?

12. What does the term *class* mean to you?
What are the most things you and other workers have in common?
What are the most important differences between workers and bosses/owners?

13. As a worker, what values and principles were most important?
What were your goals?
What rules of life did you follow?
Looking back, what motivated you in life?
What actions, events and/or experiences with other workers made the strongest impression on you?

14. What do you believe most united workers? What most tore them apart?
How were workers united outside of the plant?

15. What were the attitudes, beliefs, goals, or acts that hurt the cause of the local working class?

16. What made your community strong? What made it weak?

17. Did you view the mill, in any sense, as your property or as the property of the workers?
Why did the boss/owner have power?
How did he get it?
Did he make anything of value?
Whose mill is it?
Where does the owner's right to take what you make come from?

18. Did you believe the mills would last forever?
Were you surprised when they were closed?
What reasons were given?
Why do you think they were closed?

19. What was the most valuable aspect of your job? Why did you do the job? If for money, then what else did you get out of the job?

20. How would you describe your feelings toward other workers, the union, the job during the "good/bad" days, the 1959 strike, and the shutdown?

21. Were you a supporter or involved in Steelworkers Fight Back, RAFT (Rank-and-File Team), or the Sadlowski Campaign?

22. What literature did you read inside and outside of the plant?
Was there any knowledge/discussion of capitalism, socialism, or communism?

23. What did you and other workers complain about? Were you afraid to complain in the plant? Where did you do your complaining?

24. Who gave you orders at work? Were they always fair? If not, what did you or others do about it?
25. Did you ever try to "get your own back" (i.e., steal from the mill)?
26. Was there a lot of cursing inside the plant? Who usually got cursed?
27. Was there a lot of joking and kidding around at work? Who was involved in the horseplay? Who was the target of jokes?
28. What did you usually talk about at work?
29. What usually preceded or caused any militant shop-floor worker action?
30. What was your and other workers source of education? How did you learn things about life?

## Interviews

| Names | Company | Local Union # | Jobs | Years |
|---|---|---|---|---|
| Andrews, Kenneth | YS&T | 1462 | Brkl. Helper | 1952–1983 |
| Bator, Steve | Truscon | 2334 | Press Operator | 1942–1983 |
| Baxter, Russ | YS&T | 2163 | Bricklayer | 1953–1979 |
| Bergman, Thomas | Republic | 1331 | Shipping | 1950–1994 |
| Bodnar, George | YS&T | 1462 | Boiler Shop | 1943–1981 |
| Bruno, Robert | Republic | 1331 | Millright | 1946–1983 |
| Caban, Cayetano | YS&T | 2163 | Crane Operator | 1951–1977 |
| Calabrette, William | Republic | 1331 | Electric Weld | 1946–1985 |
| Campbell, Albert | YS&T | 2163 | Bricklayer | 1951–1980 |
| Carlini, Joe | Republic | 1331 | Welder | 1936–1983 |
| Collins, John | USS | 1330 | Open Heart | 1925–1971 |
| Costello, Jim | Republic | 1331 | Electric Weld | 1934–1976 |
| Cox, Jim | Republic | 1331 | Head Brakeman | 1945–1982 |
| Crivelli, Mario | Republic | 1331 | Stencilman | 1946–1986 |
| Cullen, Chris | Republic | 1331 | Head Hooker | 1955–1986 |
| DeChellis, Jim | YS&T | 1418 | Roller | 1939–1978 |
| Delisio, Anthony | YS&T | 1418 | Power Station | 1952–1977 |
| Delquadri, Anthony | Republic | 1331 | Electric Weld | 1943–1980 |
| Dill, Bob | YS&T | 2163 | Ladle Craneman | 1945–1977 |
| Donnely, Walter | YS&T | 1462 | Bricklayer | 1948–1980 |
| Dubos, Paul | YS&T | 2163 | Quality Control | 1951–1986 |
| Dzuroff, John | Truscon | 2334 | Press Operator | 1946–1983 |
| Esparro, Porfirio | YS&T | 2163 | Blooming Mill | 1951–1986 |
| Fattaroli, Frank | USS | 1330 | Track Gang | 1940–1978 |
| Fleming, Oscar | Republic | 1331 | Furnace Operator | 1946–1983 |
| Flora, Joe | YS&T | 1418 | Stationary Eng. | 1950–1984 |
| Floyd, Bert | YS&T | 1462 | Tractor Operator | 1942–1980 |
| Floyd, Willie | YS&T | 1462 | Ladle Liner | 1944–1980 |
| Freeman, Charles | YS&T | 2163 | Labor Gang | 1948–1980 |
| Harmon, Clifford | YS&T | none | Powerhouse Repair | 1935–1968 |
| Harp, Charlie | YS&T | 1462 | Brkl. Helper | 1952–1981 |
| Kerrick, Frank | USS | 4610 | Property Protection | 1936–1946 |
| Kotassek, Thomas | Republic | 1331 | Inspector | 1943–1986 |
| Kraynak, Tom | YS&T | 1418 | Pipe Mill | 1934–1975 |
| Luce, Merlin | USS | 1330 | Stationary Eng. | 1942–1978 |
| McGarry, John | YS&T | 2163 | Locomotive Eng. | 1953–1981 |
| Modarelli, Tony | USS | 1307 | Tractor Oper. | 1943–1980 |
| Morris, Jack | GF | 1617 | Rec. Inspector | 1947–1990 |
| Mosconi, Antigio | Republic | 1331 | Track Gang | 1948–1983 |
| Mullins, Arnett | YS&T | 2163 | Stoveman | 1946–1977 |
| Newell, Arthur | Sharon | 4418 | Electric Furnace | 1950–1990 |
| Nocera, Tony | YS&T | 1462 | Conditioning Yard | 1934–1975 |
| Occhipinti, John | Republic | 1331 | Straightner | 1936–1979 |
| Pachuta, John | Republic | 1331 | Electric Weld | 1941–1982 |
| Pajatsch, Ernest | YS&T | 1418 | Machinist | 1937–1977 |
| Pallay, John | Republic | 1331 | Coke Works | 1935–1977 |
| Papalko, George | Republic | 1331 | Shipping-Loader | 1936–1977 |
| Pellota, Tony | Sharon | 4418 | Electric Furnace | 1953–1993 |
| Perry, Tony | Republic | 1375 | Vesselman | 1948–1983 |

| Nationality | Loc. | Age at Time of Interview | Union Pos. | Organization |
|---|---|---|---|---|
| African-Amer. | Youngstown | 59 | none | |
| Slovak | Girard | 71 | Grievanceman | Solidarity |
| Scotch-Irish | Campbell | ? | President | |
| Croation | Struthers | 62 | none | |
| Slovak | Struthers | 68 | none | |
| Italian | Struthers | 64 | none | SOAR |
| Puerto Rican | Campbell | 67 | Trustee | |
| Italian | Struthers | 66 | none | SOAR |
| African-Amer. | Youngstown | 64 | Grievanceman | |
| Italian | Youngstown | 74 | President | SOAR |
| African-Amer. | Youngstown | Dec.* | none | |
| Italian | Youngstown | 77 | none | SOAR |
| Irish | Youngstown | Dec.* | none | |
| Italian | Youngstown | 67 | none | Solidarity |
| Irish | Youngstown | 60 | none | Solidarity |
| Italian | Struthers | 73 | none | |
| Italian | Struthers | 59 | none | |
| Italian | Youngstown | 68 | none | SOAR |
| Hungarian | Struthers | 73 | Vice-Pres. | |
| African-Amer. | Youngstown | 66 | none—foreman | |
| Slovak | Campbell | 61 | Appt. | |
| Slovak | Austintown | 68 | none | Solidarity |
| Puerto Rican | Campbell | 65 | none | |
| Italian | Youngstown | 72 | none | |
| African-Amer. | Youngstown | 70 | none | |
| Italian | Struthers | 61 | none | |
| African-Amer. | Youngstown | 70 | none | |
| African-Amer. | Youngstown | 65 | none | |
| African-Amer. | Youngstown | 66 | none | Solidarity |
| English | Hubbard | 90 | none—foreman | |
| African-Amer. | Youngstown | 63 | none | |
| Slovak | Youngstown | 79 | President | |
| Slovak | Struthers | 69 | none | |
| Slovak | Campbell | ? | Appt. | |
| Dutch-Irish | Youngstown | 80 | none | |
| Irish | Campbell | 66 | President | |
| Italian | Struthers | 68 | none | |
| Irish | McDonald | 65 | none | Solidarity |
| Italian | Youngstown | 70 | none | SOAR |
| African-Amer. | Youngstown | 65 | none—foreman | |
| Irish | Struthers | 67 | none | |
| Italian | Campbell | 80 | Grievanceman | |
| Italian | Struthers | 77 | none | |
| Italian | Youngstown | 73 | ? | |
| Hungarian | Struthers | 75 | none—foreman | |
| Croation | Youngstown | 77 | none | SOAR |
| Croation | Youngstown | 77 | none | |
| Italian | Hubbard | | none | |
| Italian | Hubbard | 72 | none | |

(continued)

## Interviews (*continued*)

| Names | Company | Local Union # | Jobs | Years |
|-------|---------|---------------|------|-------|
| Petrunak, Charles | YS&T | 2163 | Machine Shop | 1959–1983 |
| Phillips, John | YS&T | 2163 | Seamless Mill | 1953–1981 |
| Porrazzo, John | Republic | 1331 | Bar Mill | 1943–1985 |
| Porrazzo, George | Republic | 1331 | Tally Man | 1947–1985 |
| Ramirez, Ramon | YS&T | 1418 | Craneman | 1952–1983 |
| Ramsbottom, Brad | YS&T | 2163 | Stationary Eng. | 1946–1986 |
| Rich, James | YS&T | 1418 | Inspector | 1935–1972 |
| Rosa-Rodriquez, Ramon | Republic | 1331 | Grinder | 1951–1983 |
| Ross, Marry | Republic | none | Labor | 1943–1946 |
| Rucci, Armando | Republic | 1331 | Electric Weld | 1948–1985 |
| Ryzner, Joe | YS&T | 1331 | Bar Mill | 1936–1980 |
| Sanchez, Augustine | Republic | 1331 | Rolling Mill | 1951–1983 |
| Scocco, Mike | Republic | 1331 | Electric Weld | 1945–1983 |
| Shapiro, Sam | YS&T | 1418 | Inspector | 1952–1983 |
| Standwood, Chuck | Republic | 1331 | Clerk-Tube Mill | 1953–1984 |
| Tkach, Joseph | YS&T | 1418 | Millwright | 1920–1958 |
| Varga, John | USS | 1330 | Crane Stripper | 1941–1980 |
| Vasilchek, Mike | YS&T | 2163 | Blast Furnace | 1936–1977 |
| Visingardi, Jim | Republic | 1331 | Shipping | 1947–1985 |
| Ware, Boyd | YS&T | 1462 | Brkl. Helper | 1951–1979 |
| Woodard, Hazel | YS&T | 2163 | Burner | 1948–1981 |
| Zamary, Thomas | YS&T | 2163 | Heater | 1939–1979 |
| Zumrick, John | Republic | 1331 | Boiler Shop | 1942–1979 |

*Widows interviewed.

Interviews (*continued*)

| Nationality | Loc. | Age at Time of Interview | Union Pos. | Organization |
|---|---|---|---|---|
| Slovak | Struthers | 62 | none—foreman | |
| Russian | New Castle | 74 | none | |
| Italian | Youngstown | 69 | none | SOAR |
| Italian | Youngstown | 67 | Appt. | SOAR |
| Puerto Rican | Youngstown | 68 | none | |
| English | Struthers | 68 | Vice-Pres. | |
| Italian | Campbell | 76 | none | |
| Puerto Rican | Youngstown | 70 | none | |
| Croation | Youngstown | 80 | none | SOAR |
| Italian | Struthers | 66 | none | SOAR |
| German | Liberty Twp. | 75 | none | SOAR |
| Puerto Rican | Youngstown | 65 | none | Solidarity |
| Italian | Youngstown | 71 | Inside Guard | SOAR |
| Jewish | Youngstown | | none | Solidarity |
| Irish | Youngstown | 66 | none | Solidarity |
| Slovak | Struthers | Dec.* | none | |
| Hungarian | Struthers | 73 | none | |
| Slovak | Youngstown | 78 | none—foreman | |
| Italian | Struthers | 66 | none | |
| African-Amer. | Youngstown | 64 | none | |
| African-Amer. | Youngstown | 77 | none | |
| Croation | Struthers | Dec.* | none | |
| Slovak | Youngstown | 76 | none | SOAR |

# Notes

## Introduction: His Silence Broken

1. Godfrey Hodgson, *America in Our Time* (New York: Vintage Books, 1976), 82. While any number of post–World War II writings could be used to demonstrate an unfolding "liberal consensus," a short, multidisciplinary list would certainly include the following: Daniel Bell, *The End of Ideology* (Glencoe, Ill.: Free Press, 1960); Clark Kerr, Charles M. Myers, John Dunlop, Frederick H. Arbison, *Industrialism and Industrial Man* (London: Oxford University Press, 1962); Seymour Martin Lipset, *Political Man: The Social Bases of Politics* (New York: Anchor Books, 1960); John Kenneth Galbraith, *The Affluent Society* (Boston: Houghton Mifflin, 1958); Peter Drucker, *The New Society: The Anatomy of the Industrial Order* (New York: Harper, 1950); Arthur M. Schlesinger Jr., *The Vital Center: The Politics of Freedom* (Boston: Houghton Mifflin, 1949); and Samuel Eliot Morison and Henry Steele Commager, *The Growth of the American Republic* (New York: Oxford University Press, 1969).

2. *Militant*, 23 January 1956.

3. For two classic surveys that reach different conclusions on what the working class thinks, see John Goldthorpe et al., *The Affluent Worker in the Class Structure* (Cambridge: Cambridge University Press, 1969), and Robert Lane, *Political Ideology: Why the American Common Man Believes What He Does* (Glencoe: Free Press, 1972). An untheorized, yet unparalleled subjective description of the working class is Studs Terkel's *Working* (New York: Ballantine Books, 1974). A relevant look at working-class hesitancy can be found in Thomas G. Fuechtmann, *Steeples and Stacks: Religion and Steel Crisis in Youngstown* (Cambridge: Cambridge University Press, 1989). Philip W. Nyden, *Steelworkers Rank-And-File: The Political Economy of a Union Movement* (New York: Praeger Press, 1984), offers a critical examination of the failure of dissident labor movements in the 1970s. A good overview of the consciousness

and "identity" debates within the labor history community is Michael Kazin's "Struggling with Class Struggle: Marxism and the Search for a Synthesis of U.S. Labor History," *Labor History* 28 (1987), 497–514.

4. Perhaps the most influential critique of the "thesis of embourgeoisement" was *The Affluent Worker*, by John Goldthorpe, David Lockwood, Frank Bechhoffer, and Gennifer Platt. In brief, while workers were found to have experienced profound changes in their lives, "such changes need not betoken the adoption of specifically middle-class models of sociability," 158, and "assimilation into middle-class society is neither in process nor, in the main, a desired objective," 157. Additional challenges came from David Lockwood, "Sources of Variation in Working-Class Images of Society," in *Classes, Power, and Conflict*, ed. Anthony Giddens and David Held (Berkeley: University of California Press, 1982); Frank Parkin, *Class Inequality and Political Order: Social Stratification in Capitalist and Communist Societies* (New York: Praeger Press, 1972); and Frank Reisman and S. M. Miller, "Are Workers Middle Class?" *Dissent* 8 (Fall 1961).

5. For a crisp survey of the various social and intellectual contributions to a post–World War II "liberal consensus" see Hodgson's *America*, 67–98.

6. Ibid., 76.

7. David Halle, *America's Working Man* (Chicago: University of Chicago Press, 1984), and Michael Frisch and Milton Rogovin, *Portraits in Steel* (Ithaca: Cornell University Press, 1993).

8. Rick Fantasia, *Cultures of Solidarity: Consciousness, Action, and Contemporary American Workers* (Berkeley: University of California Press, 1988), 4.

9. See Herbert Gutman's collected essays in *Power and Culture: Essays on the American Working Class*, ed. Ira Berlin (New York: Pantheon Books, 1987).

10. John Legget, "Economic Insecurity and Working-Class Consciousness," *American Sociological Review* 29 (1968), 226–234.

11. Joseph Schumpeter, *Imperialism and Social Classes*, trans. Heinz Norden (New York: Augustus M. Kelly, 1951), 76–77. Class has also been demonstrated to be a medium for personal attitudes, beliefs, images, and behaviors. Distinct classes have been shown to reflect dichotomous opinions on societal issues, and where political issues are the focus class divisions have been the most important. See Reeve Vanneman and Fred C. Pampel, "The American Perception of Class and Status," *American Sociological Review* 43, no. 3 (1977), 422–438.

12. "Letter to Henry Champion," in *Struthers: Prologue-Epilogue*, Struthers High School. Ohio Department of Education, ESEA Grant Title III, Project No. 45-74-527-2 (1975), 29.

13. *Industry and Commerce in Youngstown*, Youngstown Committee of the Ohio Sesquicentennial Commission, 1803–1953 (Youngstown, 1953), 18.

14. Ibid., 19.

15. Other initial owners included George Fordyce, James Campbell, E. L. Ford and William Wilkoff. The name of the sixth person has inexplicably been dropped, although I suspect it was one of the Tod family members. Charles Carr, "The First Fifteen Years of the Youngstown Sheet & Tube Co.," *Youngstown Vindicator*, 21 November 1915.

16. Carr, "The First Fifteen Years," 2. Along with James Campbell, the Tods, Wicks and Stambaughs accounted for most of Youngstown's early-twentieth-

century venture capital, and the names remain to this day a symbol of local royalty. These names also appear on stadiums, streets, college and medical buildings, playhouses, and parks.

17. To be exact, thirty-five companies in six states were merged to create Republic Iron and Steel. The enlarged company first incorporated in 1899. While not listed as a principle owner, Myron Wick was elected first vice-president and chairman of the company's executive committee. Colonel George Wick also held a vice-president position and James Campbell worked as a company district superintendent before leaving to organize YS&T. Charles Carr, "Republic Iron & Steel Co's Great Plant Rose from Small," *Youngstown Vindicator*, 5 December 1915.

18. *Industry and Commerce in Youngstown*, 18–22.

19. Ibid.

20. The reference to the Ruhr Valley was apparently part of an early encyclopedic description of the Mahoning Valley. See "10 Years After: Black Monday Signaled Demise of Valley Steel," *Youngstown Vindicator*, 13 September 1987.

21. Quoted from Staughton Lynd's *The Fight against Shutdowns: Youngstown's Steel Mill Closings* (San Pedro, Calif.: Singlejack, 1983), 20–21.

22. The story of Youngstown's deindustrialization is told by Lynd in *The Fight against Shutdowns*.

23. The only remaining steelworkers employed for one of the three principal companies are working at LTV Steel's Tubular Products Plant in Youngstown ("The New Strength in Steel," *Reporter*, an LTV Steel publication, May 1993). Ling Temco Vought (LTV) purchased Republic Steel in 1984 and merged it with Jones and Laughlin. Claiming staggering losses to their steel division, LTV Corporation then entered bankruptcy in 1986 and remained under court protection until the summer of 1993 ("In the Beginning," *Youngstown Vindicator*, 13 September 1987). There has been some steel production outside the major three. In 1982, Hunt Steel began operating a seamless mill on the premises of YS&T's old Brier Hill location. However, it entered bankruptcy in 1984 and was purchased by mini-mill specialist North Star Steel ("In the Beginning," *Youngstown Vindicator*, 13 September 1987).

24. For a listing of companies, plants, and dates see Father William T. Hogan, S.J., *Steel in Crisis* (Pittsburgh: Steel Communities Coalition, 1977), Exhibit 1.

25. In 1953, employment for production and maintenance stood at over 571,000. See Tom DuBois, "Steel: Past the Crossroads," *Labor Research Review*, Midwest Center for Labor Research (Winter 1983), 14.

26. John P. Hoerr, *And the Wolf Finally Came: The Decline of the American Steel Industry* (Pittsburgh: University of Pittsburgh Press, 1988), 17.

27. Jack Metzgar, "Would Wage Concessions Help the Steel Industry?" *Labor Research Review*, Midwest Center for Labor Research (Winter 1983), 30.

28. Apparently Roderick and other corporate officers were also in the moneymaking business. Roderick took $783,750 in compensation out of U.S. Steel (USS), while LTV Steel Chairman Paul Thayer helped himself to a bit more, $1,163,622. See Metzgar, "Would Wage Concessions Help the Steel Industry?" 26.

29. In 1982 USS spent $6 billion to buy Marathon Oil. In 1985 the company dropped the last S and substituted its stock exchange symbol, an X. See Hoerr, *And the Wolf Finally Came*, 521.

30. In 1920 Italians constituted 16 percent of the foreign-born population in Youngstown. Slovakians, Croations, and Poles accounted for 27 percent; Hungarians, Rumanians and Russians, 18 percent; and Austrians, 9 percent. See U.S. Department of Commerce, Bureau of the Census, "Youngstown, Ohio, Census of Foreign Born Population, 1900–1960," in *United States Census of the Population, 1960* (Washington, D.C.: GPO, 1961).

31. Data taken from a table in Gutman, *Power and Culture*, 388.

32. Sharon Steel was centered in neighboring Shenango Valley, Pennsylvania. The company was originally formed as Sharon Steel Hoop in 1900 and later expanded through local acquisitions. Sharon Steel operated a plant in the city of Loweville, lying between Struthers and Campbell, where a number of the workers that I interviewed were employed. See *Industry and Commerce in Youngstown*, 20.

33. Descriptions of SOAR are taken from a USWA publication for retirees called *The Oldtimer* and from personal observations at SOAR meetings in the summer of 1993.

34. A history of Solidarity is printed in the Youngstown Workers Club monthly newsletter, *Impact: The Rank & File Newsletter* 1, no. 4 (July 1993).

35. Data printed in Terry F. Buss and F. Stevens Redburn, *Reemployment after a Shutdown: The Youngstown Steel Mill Closing, 1977–1985* (Youngstown, Ohio: Center for Urban Studies, Youngstown State University, 1986).

36. See Mike Davis, *Prisoners of the American Dream* (London: Verso Press, 1986) and "The Barren Marriage of American Labor," *New Left Review* 124 (November/December 1980), 43–84. Historically the absence of socialism has been proof of political capitulation to capital. For an unconventional critique of American labor's alleged lack of radicalism, see Eric Foner, "Why Is There No Socialism in the United States?" *History Workshop* 17 (1984), 57–80.

37. Ira Katznelson, "Working-Class Formation: Constructing Cases and Comparisons," in *Working-Class Formation: Nineteenth-Century Patterns in Western Europe and the United States*, ed. Katznelson and Aristide R. Zolberg (Princeton Press, 1986), 3, 18.

## Chapter 1: Steel-Paved Streets

1. Perhaps one of the most impressive studies of contemporary workers, yet representative of the spatial-duality theme, is David Halle's *America's Working Man* (Chicago: University of Chicago Press, 1984). The crux of Halle's findings was that blue-collar workers possessed a dual consciousness built around a composite definition of the term "working man." Halle identified a complex workers' view of the class structure by pointing out that workers inhabited two contradictory worlds, one at work and one outside of work.

2. Here is a complete list of Mahoning Valley steel mills: Youngstown Sheet and Tube (4 plants), Republic Steel (2 plants), United States Steel (2 plants), Sharon Steel Hoop (1 plant), Copperweld Steel (1 plant). Data from Western Re-

serve Economic Development Agency, in Thomas G. Fuechtmann, *Steeples and Stacks: Religion and Steel Crisis in Youngstown* (Cambridge: Cambridge University Press, 1989), 15.

3. *Industry and Commerce in Youngstown*, Youngstown Committee of the Ohio Sesquicentennial Commission, 1803–1953 (Youngstown, 1953), 22.

4. Material cited in Fuechtmann, *Steeples and Stacks*, from *United States Department of Commerce, County Business Patterns* (1975), 21. The following is a 1975 list of principal non-steelmaking manufacturing firms with over a thousand employees:

|                                    | *Employment* |
| ---------------------------------- | ------------ |
| General Motors                     | 17,474       |
| Wean United                        | 3,545        |
| General Electric                   | 2,431        |
| General Fireproofing               | 2,250        |
| General American Transportation    | 2,146        |
| American Welding                   | 1,109        |
| Reactive Metals, Inc.              | 1,082        |
| Youngstown Steel Door              | 1,080        |
| Commercial Shearing                | 1,023        |
| North American Rockwell            | 1,008        |

Data cited in Fuechtmann, *Steeples and Stacks,*
from Youngstown Chamber of Commerce.

5. Fuechtmann, *Steeples and Stacks*, 13. In 1926, East Youngstown was officially incorporated as the city of Campbell.

6. Information collected from Mahoning County Department of Deeds and Records, Mahoning County Courthouse, Youngstown, Ohio.

7. Mahoning County Tax Records, Mahoning County Courthouse, Department of Tax Assessor, Youngstown, Ohio; and *Youngstown City Directory*, 1952.

8. *Youngstown City Directory*, 1952.

9. The manufacturing cities of Campbell and Loweville would have looked similar. Campbell was home to YS&T, and Loweville home to Sharon Steel.

10. These quotes were so widely shared as to be practically generic. Each of the workers I interviewed either used these exact words or words that were very similar.

11. Fuechtmann, *Steeples and Stacks*, 18, 27.

12. *Struthers Echo Class Yearbook*, 1940.

13. U.S. Department of Commerce, Bureau of the Census, "Labor Force Characteristics of the General Population, Youngstown, Ohio," in *United States Census of Population, 1960* (Washington, D.C.: GPO, 1961); and *Youngstown City Directory*, 1952.

14. Abstract of property deed to Mike Vasilchek's home.

15. Housing data collected from Mahoning County Tax Records, 1993.

16. Quoted from Fuechtmann's reference in *Steeples and Stacks*, 25, to Christine Franz-Goldman, "Socioeconomic Costs and Benefits of the Community-Worker

Plan to the Youngstown-Warren SMSA," in *Youngstown Demonstration Planning Project: Final Report* (Washington, D.C.: National Center for Economic Alternatives, 1978).

17. Obituaries, *Youngstown Vindicator*, 15 June 1993.

18. *Youngstown Vindicator*, 10 June 1993.

19. Ibid., 23 June 1993.

20. My argument here inverts the claims of many urban sociologists who would give priority to ethnic composition in explaining working-class neighborhood development. For a good review of the neighborhood/ethnicity relationship see Kathleen Neils Cozen, "Immigrants, Immigrant Neighborhoods, and Ethnic Identity: Historical Issues," *Journal of American History* 66, no. 3 (December 1979), 603–615.

21. Quoted in Cozen, "Immigrants, Immigrant Neighborhoods, and Ethnic Identity," 606.

22. *Youngstown City Directory*, 1952.

23. *Struthers Journal*, 26 April 1937 and 1 March 1961.

24. Ibid., 26 February 1937.

25. This information was presented to me from conversations with numerous workers and their spouses.

26. The class-building potential of integrated neighborhoods is stressed by Ira Katznelson's reference to Craig Calhoun's essay "Class, Place and Industrial Revolution," in *Marxism and the City*, ed. Ira Katznelson and Aristide R. Zolberg (Oxford: Clarendon Press, 1992), 267–273.

27. George D. Beelen, "The Negro in Youngstown: Growth of a Ghetto" (paper, Kent State University, 1967), 1–17.

28. Marvin Weinstock, president of Local Union 1330, and a black fellow union member attempted in the late 1940s to enter an all-white public swimming pool. Weinstock recalls that once they got into the water, both workers were quickly set upon by a large number of young white men. Weinstock felt that his presence actually angered the crowd more than the presence of his black co-worker. The incident was written up in the following day's *Youngstown Vindicator* and the paper gave particular attention to an announcement by the local Socialist Workers Party that a rally would be held at the pool (Marvin Weinstock, transcripts of interview Ohio Historical Society Oral History Project, Youngstown Historical Center of Industry and Labor, Youngstown, Ohio, 28 August 1991).

29. U.S. Department of Commerce, Bureau of the Census, "Number and Distribution of the Population by Race, Sex, Median Age, and Census Tracts, Youngstown, Ohio, 1960," in *United States Census of Population, 1960* (Washington, D.C.: GPO, 1961).

30. U.S. Department of Commerce, Bureau of the Census, "General Characteristics of the Population By Census Tracts, Youngstown, Ohio," in *United States Census of Population, 1960* (Washington, D.C.: GPO, 1961).

31. Thirty-six percent of the black population resided on the east side (tracts 1–9), 28 percent lived on the south side (tracts 10–25), and 33 percent on the north side (tracts 33–44). Data taken from U.S. Department of Commerce, "Number and Distribution."

32. The quote is from YS&T worker John Barbero, printed in *Rank-and-File: Personal Histories by Working-Class Organizers*, ed. Alice Lynd and Staughton Lynd (New York: Monthly Review Press, 1988), 262.

33. U.S. Department of Commerce, "Labor Force Characteristics."

34. *Youngstown City Directory*, 1952.

35. *Cradle of Steel Program*, 1948.

36. Ibid.

## Chapter 2: Santa Claus Was a Steelworker

1. Workers not only referred to themselves in this way, but local merchants, clerics, and politicians regularly prefaced any remarks or addresses to steelworkers with a respectful salutation. One black worker even pointed out that in his neighborhood "steelworkers were considered elites" (James Davis, transcript of interview, 22 April 1991, Youngstown Historical Center of Industry and Labor, Youngstown, Ohio).

2. The process also worked in the opposite direction. Childhood friendships and school-grown experiences were customarily followed by common workplace settings.

3. Parishioners at St. Nicholas comprised roughly 65 percent of the city's population, and the parish was the largest in the Youngstown Diocese (*Spirit*, Saint Nicholas Parish Weekly, Youngstown Diocese, 14 March 1965).

4. This Mass was held before the official 6:45 A.M. start of Mass (Church Bulletin, Saint Nicholas Parish, Youngstown Diocese, 31 January 1965). Jim DeChellis, interviewed by author, tape recording, Struthers, Ohio, 3 June 1993.

5. Baptismal Register, Saint Nicholas Parish, Youngstown Diocese, 16 August 1944–26 December 1954.

6. Robert Bruno, interview by author, tape recording, Struthers, Ohio, 30 August 1993; and Joe Carlini, interview by author, taped recording, Struthers, Ohio, 9 June 1993. A brief sample of the organized and promoted activities at St. Nicholas in 1977 includes: "Card Party, Workshop on Human Rights, Health-O-Rama, Chicken Dinner, Senior Citizen & Handicapped Transportation, Bingo, Harvest Moon Dance, League of Women Voters meeting, and Men's and Women's Choir" (Church Bulletin, 16 October 1977).

7. DeChellis, interview.

8. One month before the start of the 1959 strike, parishioners offered $531.25 during one day of Sunday Masses (Church Bulletin, 19–26 April and 3–10 May 1959).

9. Church Bulletin, 25 September 1977. The church continued to stress the mill closings and the community's response in subsequent bulletins. Announcements featured phrases like "Concerned—Yes—Discouraged—No!" and "Pray Together For Each Other." Most messages called for a dignified rational reaction and discouraged panic, and "sudden drastic actions." The church appealed for prayer and confidence in God. Curiously, while the bulletins referred to the "Steel Mill Crisis" and asked for faith in the wisdom "of government and industrial" leaders, they never once used the word "workers." It appears from

published records that St. Nicholas Church did not act to mobilize working-class sentiments against the shutdowns (Church Bulletins, 25 September 1977–16 October 1977).

10. The announcement for the Mass encouraged parishioners to "immediately turn to God for help and guidance," and, "like the Israelites of old," people had to "come to God's Temple of the Father." The message concluded with a reading from 1 Corinthians 12:26: "If one member suffers, all the members suffer with it" (Church Bulletin, 25 September 1977).

11. Thomas G. Fuechtmann, *Steeples and Stacks: Religion and Steel Crisis in Youngstown* (Cambridge: Cambridge University Press, 1989), 138.

12. The full text is printed in Fuechtmann, *Steeples and Stacks*, 137–145.

13. One curious research note suggests that this letter was in fact read and viewed as meaningful to steelworkers. In interviewing George Bodnar (tape recording, Struthers, Ohio, 16 June 1993) I came across a copy of the letter and an information sheet on the coalition in a stack of his personal papers. Bodnar's papers were titled "A Response to the Mahoning Valley Steel Crisis" and "Ecumenical Coalition of the Mahoning Valley: Background and Future Projections." Both documents were published by the coalition in 1977. The latter piece included the following biblical verse as the basis for the coalition's mission:

> "The spirit of the Lord Yahweh has been given
> to me, for Yahweh has anointed me.
> He has sent me to bring good news to the poor,
> to bind up the hearts that are broken;
> to proclaim liberty to the captives,
> freedom to those in prison;
> to proclaim a year of favor for Yahweh,
> a day of vengeance for our God. . .".
> (Isaiah 61:1–2, Jerusalem Bible)

14. Allan Pred argued that the split between work and home presented workers with a choice between "free-time projects defined by occupationally and ethnically oriented institutions" that would in all likelihood be "mutually exclusive." In the case of Youngstown-area steelworkers, associations free of occupational characteristics may have been impossible. See Allan Pred, "Production, Family, and Free-Time Projects: A Time-Geographic Perspective on the Individual and Societal Change in Nineteenth-Century U.S. Cities," *Journal of Historical Geography* 7 (January 1981), 3–36.

15. U.S. Department of Commerce, Bureau of the Census, "General Characteristics of the Population by Census Tracts, Youngstown, Ohio" (Washington, D.C.: GPO, 1961).

16. *100th Anniversary Program*, Italian-American Club, 1976.

17. Local history of the city of Loweville, titled "History of Loweville and the First Italian Immigrants That Arrived," by Robert Delisio (1975), 27.

18. Ibid., 12.

19. *Struthers (Ohio) Journal*, 9 February 1956.

20. Ibid.

21. Ibid., 1 March 1956.

22. Ibid., 11 October 1956.

23. Ibid., 29 March 1956.

24. Ibid., 4 October 1956.

25. By 1971, approximately 12 percent of YS&T's workforce was Puerto Rican (*Trade and Craft Utilization Analysis*, Youngstown District Operations, Youngstown Sheet and Tube, 2 August 1979).

26. U.S. Department of Commerce, Bureau of the Census, "Number and Distribution of the Population by Race, Sex, Median Age, and Census Tracts, Youngstown, Ohio, 1960," in *United States Census Of Population: 1960* (Washington, D.C.: GPO, 1961).

27. Ibid., Census tracts E7–E9 and N37–N39.

28. Steven Riesman's *City Games: The Evolution of American Urban Society and the Rise of Sports* (Urbana: University of Illinois Press, 1989) demonstrates the close connection between athletics and working-class identity in urban areas.

29. *Struthers Journal*, 20 September 1956 and 2 July 1959.

30. Ibid., 30 April 1959.

31. Ibid., 3 July 1956.

32. Ibid., 10 July 1956.

33. Ibid., 22 March 1956.

34. Steve Bator and John Dzuroff, interview by author, tape recording, Austintown, Ohio, 12 July 1993.

35. John Zumrick, interview by author, tape recording, Youngstown, Ohio, 4 June 1993.

36. Oscar Flemming, Youngstown, Ohio, 9 July 1993; Joe Flora, Struthers, Ohio, 10 June 1993; and Mike Scocca, Struthers, Ohio, 14 June 1993; George Papalko, Youngstown, Ohio, 2 June 1993; and Annie Frattaroli, Youngstown, Ohio, 8 June 1993, all interviews by author, tape recordings.

37. Information taken from Local Union 1331 sample amusement park tickets and *AFL-CIO 1331 Local News*, 29 October 1969, part of George Porrazzo's personal collection.

38. *Struthers Journal*, 3 January 1957.

39. *Ohio Works Organizer* (USWA Local 1330), 14 December 1956.

40. Ibid., 4 November 1961.

41. Ibid., 1 September 1952.

42. *Struthers Journal*, 20 August 1937.

43. *Ohio Works Organizer*, 20 September 1946.

44. Ibid., 29 November 1946, 5 February 1948, and 27 April 1959.

45. *Struthers Journal*, 17 February 1955 and 14 April 1955.

46. Ibid., 9 February 1956.

47. *Ohio Works Organizer*, 14 December 1956.

48. *Youngstown Vindicator*, 19 July 1959.

49. Ibid., 12 July 1959. Local 2163 was still sponsoring city baseball and union softball teams in the mid-1970s (letter from Local Union 2163 president Russ Baxter to membership, in Bob Dill's personal records).

50. *Ohio Works Organizer*, 14 October 1946.

51. Ibid., 1 April 1948.

52. Ibid., 14 October 1946, 27 April 1959, and 4 November 1961. Also see the *Brier Hill Unionist* (Local 1418 of YS&T), issues from 1971 to 1979.

## Chapter 3: Fried Onions and Steel

1. Information taken from *Job Description and Classification Manual,* United Steelworkers of America, AFL-CIO and Coordinating Committee Steel Companies, 1 January 1963, 15.

2. *Job Description and Classification Manual,* Job no. 60.

3. *Job Description and Classification Manual,* Job no. 456. My father did this job for the bulk of his career at Republic Steel. When he retired in 1985, the job earned a Class 18 rating. The working conditions were still bad.

4. Robert Bruno, interview by author, tape recording, Struthers, Ohio, 30 August 1993.

5. Material taken from personal work documents of YS&T employee Tom Kraynak.

6. Ira Katznelson, *Marxism and the City* (New York: Oxford University Press, 1992), 105.

7. Red Delquadri, interview by author, tape recording, Youngstown, Ohio, 7 June 1993.

8. Bill Calabrette, interview by author, tape recording, Struthers, Ohio, 4 June 1993; and Jim DeChellis, interview by author, tape recording, Struthers, Ohio, 3 June 1993.

9. Tom Kotasek, interview by author, tape recording, Struthers, Ohio, 11 June 1993.

10. John Varga, interview by author, tape recording, Struthers, Ohio, 2 June 1993.

11. Russ Baxter was a five-term president of USWA Local 2163, and while most of the workers he fought for were "guilty of some infraction," he believed it was necessary to always defend them, even if it meant lying (Russ Baxter, interview by author, tape recording, Campbell, Ohio, 13 June 1993).

12. Oscar Flemming, interview by author, tape recording, Youngstown, Ohio, 9 July 1993; and Mario Crivelli, interviews by author, tape recording, New Bedford, Pennsylvania, 7 July 1993. Another white Republic worker, Tom Kotasek, felt that Pete Starks was the "most aggressive grievance man in the mill." My father also spoke very positively of Stark's efforts on behalf of all workers (Kotasek, interview; and Bruno, interview).

13. *Trade and Craft Utilization Analysis and Goals and Timetables, for Brier Hill and Campbell Plants,* 2 August 1979; *Youngstown Sheet and Tube Company—Youngstown District, Report 1615, Union Lodge Report,* 27 January 1973.

14. Danny Thomas was president of Local 1462 at YS&T's Brier Hill plant and remembered that prior to the late 1950s the company had maintained separate bathrooms, shower rooms, and changing areas (Danny Thomas, transcript of interview, 1 August 1974, Ohio Historical Society Oral History Project, Youngstown Center of Industry and Labor, Youngstown, Ohio). Al Campbell felt that after the civil rights laws were passed the company began to treat black

workers with greater respect. While he was critical of union racism, Campbell credited both the union and the federal government with the overall improvement in race relations (Albert Campbell, interview by author, tape recording, Youngstown, Ohio, 17 June 1993).

15. Quoted in Alice Lynd and Staughton Lynd, eds., *Rank-And-File: Personal Histories by Working-Class Organizers* (New York: Monthly Review Press, 1998), 267.

16. The agreement, known as the "Consent Decree," settled a civil rights suit brought by minority and female steelworkers against nine steel companies, including those operating in Youngstown, and the USWA. Among other features, the decree established job promotional ladders on the basis of plant-wide seniority and required that the steel companies (found to be more "guilty" of discrimination than the union) create a "back pay" compensation pool for workers previously discriminated against. Payment to each "affected employee" was to be no less than $250 (*Consent Decree I*, United States Court for the Northern District of Alabama, Southern Division, Civil Action No. 74P339, 12 April 1974).

17. James Davis, transcript of interview, 22 April 1991, Ohio Historical Society, Youngstown Historical Center of Industry and Labor, Youngstown, Ohio.

18. Brad Ramsbottom, interview by author, tape recording, Struthers, Ohio, 12 June 1993. Brad did not remember a single black worker in his department until the Consent Decree in 1974.

19. Here is a sample of last names from the open hearth seniority list at the Campbell Plant of YS&T: Dworak, Flasck, Nettles, Knapick, Skripac, Koehler, Shevitz, Hushar, and Petko (*Campbell Open Hearth Seniority List, 1970*).

20. Getting around sometimes meant driving around. Bert Floyd was a black forklift driver at YS&T who went all over the plant bringing an assortment of materials. While his best friends were black, he claimed to personally know nearly every man in the plant (Bert Floyd, interview by author, tape recording, Youngstown, Ohio, 5 July 1993).

21. This condition was prevalent in the open heart, particularly during a strip down of the old brick. After about a hundred heats, a furnace needed to be shut down and the old brick lining torn out and re-bricked. These jobs were setup and cleaned up by laborers and semi-skilled workers.

22. There was some disagreement among white and black workers about the degree of separation. YS&T employee Kenneth Andrews spoke of white workers keeping their "tools in one room and eating in separate places." But Andrews aside, most workers felt that the physical distance kept between races was minimal (Kenneth Andrews, interview by author, tape recording, Youngstown, Ohio, 30 June 1993). However, at YS&T's Brier Hill Plant, there was a workers' bathroom divided by a makeshift wall. The company had constructed it during the war to ostensibly provide some degree of privacy, but workers understood that the real reason was to prevent black and white workers from washing and changing in the same area. The wall stood during the 1940s, then it was ceremoniously torn down by a group of white workers, led by Local 1462 president Danny Thomas. As Thomas succinctly put it, "We picked up sledge hammers and we smashed down the walls" (Danny Thomas, transcript of interview, 1 August 1974, Ohio Historical Society Oral History Project, Youngstown Historical Center of Industry and Labor, Youngstown, Ohio, 17).

23. John Occhipinti, interview by author, tape recording, Struthers, Ohio, 10 June 1993.

24. Letter from Frank J. Weldele, Ph.D., Director of Audiology, St. Elizabeth Hospital Medical Center, to Jones and Laughlin Steel Corporation, Industrial Relations Department, 23 February 1981 (Anthony Delisio's personal records).

25. "Memorandum of Agreement," between Local 1418 and YS&T, 14 April 1982 (Anthony Delisio's personal records). Also, the "Rule-of-65" allowed steelworkers to retire with a full pension if their age and years of service added up to sixty-five.

26. "The Wildcat over Tony's death," in Alice Lynd and Staughton Lynd, eds., *We Are the Union: The Story of Ed Mann*, distributed by Solidarity USA, 36.

27. Anthony Delisio, interview by author, tape recording, Struthers, Ohio, 3 June 1993.

## Chapter 4: Making "Good Money" on Time and Credit

1. Daniel Bell, *The End of Ideology* (Glencoe, Ill.: Free Press, 1960); Ralf Dahrendorf, *Class and Class Conflict in Industrial Society* (Stanford: Stanford University Press, 1959); and Clark Kerr, "Industrial Conflict and Its Mediation," *American Journal of Sociology* 60, no. 3 (November 1954).

2. James O'Toole et al., *Work in America* (Cambridge: MIT Press, 1973), summary, xv.

3. Ibid.

4. Ibid., xvi.

5. Richard Parker, *The Myth of the Middle Class: Notes on Affluence and Equality* (New York: Liveright, 1972), introduction. Gabriel Kolko posits a similar inequality in class distinction in his *Wealth and Power in America: An Analysis of Social Class and Income Distribution* (New York: Praeger Press, 1962).

6. Parker, *Myth of The Middle Class*, 7.

7. Ibid., 8.

8. Ibid., 9.

9. Ibid.

10. Andrew Levison, *Working-Class Majority* (New York: Penguin Books, 1975), 13.

11. Ibid., 31–32.

12. Ibid., 20.

13. Ibid., 37–38. Richard Hamilton, in *Class and Politics in the United States* (New York: John Wiley, 1972), refers to the idea of a uniform, universal-class domestic setting as one of many misguided and exaggerated "look-alike" theses, 163.

14. Mark McColloch, "Consolidating Industrial Citizenship: The USWA at War and Peace, 1939–46," in *Forging a Union of Steel: Philip Murray, SWOC, and the United Steelworkers*, ed. Paul Clark, Peter Gottlieb and Donald Kennedy (Ithaca, N.Y.: ILR Press, 1987), 62–69.

15. McColloch, "Consolidating Industrial Citizenship," 63.

16. Bureau of Labor statistical data reported in the *Ohio Works Organizer* (USWA Local 1330), 8 February 1952.

17. *Ohio Works Organizer*. The Consumer Price Index from June 1950 to March 1951 was up by 14.5 percent, while steel wages increased by 7.6 percent.

18. John P. Hoerr, *And the Wolf Finally Came: The Decline of the American Steel Industry* (Pittsburgh: University of Pittsburgh Press, 1988), 72.

19. Data computed from testimony by workers about their "best years" and estimated tax payments taken from George Porrazzo's financial records. For example, in 1974 Porrazzo earned $20,676 but paid $5,693.49 (28 percent) in taxes.

20. George Porrazzo's personal work journals from 1966 to 1971.

21. Red Delquadri's personal financial records.

22. The Delquadris' normal monthly expenses for "staples" in the early 1960s amounted to $115. Additional costs included installment payments of $26.25 and $13.62 to Stambaugh Thompson, a hardware store (household budget for January and February of 1961, in Red Delquadri's personal records). Prior to company subsidized health insurance in 1960, even family medical bills were sometimes left partially unpaid. In 1955, Delquadri made $20 payments to work off a $1,200 balance (medical receipts in Red Delquadri's personal records).

23. Joe Flora's personal financial records.

24. Earnings data taken from pay stubs covering 1950 to 1970 (Armando Rucci's personal records).

25. Income tax records from Joe Flora's personal records.

26. Armando Rucci's personal records.

27. Red Delquadri's savings account bank book, 1945–1975.

28. Robert Bruno, interview by author, tape recording, Struthers, Ohio, 30 August 1993.

29. Data taken from pay notebook covering the period from 3 March 1971 to 24 May 1971 (Tom Kraynak's personal records).

30. George Porrazzo's personal financial records, including pay stubs, covering his entire working career. The weeks referred to in the text are from 3 January 1974 (gross pay $330.66) to 17 January 1974 (gross pay $591.39).

31. Data taken from pay stubs covering 1950 to 1970 (Armando Rucci's personal records).

32. "Coke Plant Sequence" taken from Albert Campbell's personal records.

33. According to Charley Harp and Boyd Ware, 80 percent of the helpers at YS&T's Campbell Works were black. Charley Harp, interview by author, tape recording, Campbell, Ohio, 2 July 1993; and Boyd Ware, interview by author, tape recording, Campbell, Ohio, 3 July 1993.

34. Carlini served five terms as Local 1331 president from 1964 to 1967 and from 1973 to 1985. Joe Carlini, interview by author, tape recording, Struthers, Ohio, 9 June 1993.

35. Augustine Izzo, Local 1462, transcript of interview, Ohio Historical Society Oral History Project, Youngstown Historical Center of Industry and Labor, Youngstown, Ohio, 26 March 1991.

36. A curious and unexpected aspect of my interviews was that most white workers spoke positively about their earnings. Many said that they "made good

money," that their earnings "made a good life," and that they "earned a good living." While few believed they were fairly compensated, most were not angry about the level of compensation they received. It was not, however, until I examined financial statements and pay stubs that I realized that "good pay" amounted to considerable overtime, shift differential, and incentive pay. Without bonus money, workers struggled financially, and for many workers base pay was all they earned.

37. Michael Lebowitz, *Beyond Capital: Marx's Political Economy of the Working Class* (New York: St. Martin's Press, 1992), 67.

38. The Industrial Department of the AFL-CIO submitted to local steel unions copies of guidelines and instructions for confronting company imposed time-motion studies. The material was to be used to challenge company efforts to determine work rates and to expose the common ways foremen would deceive engineers. Data taken from "Methods of Deception," by Bertram Gottlieb (1964), in George Porrazzo's personal collection.

39. Material taken from tax records (Joe Flora's personal records).

40. The $26 figure put out by the American Iron and Steel Institute actually represents wages and all benefits of active workers in addition to all benefits paid to laid-off workers. The actual "money wage" in 1981 was $12.09, but adjusted for inflation it was only $4.44 per hour. See Tom Dubois, "The Myth of the $26-an-Hour Steelworker," *Labor Research Review: The Crisis In Steel* (Winter 1983), 52–53.

41. Strikes occurred in 1946, 1940, 1952, 1956, 1959 and 1962 . See Paul Clark, Peter Gottlieb, and Donald Kennedy, appendix, in *Forging A Nation Of Steel*.

42. *Youngstown Vindicator*, 8 August 1959.

43. The food aid was broken down in the following categories: flour, 26,530 pounds; dried milk, 6,448 pounds; and eggs, 4,760 pounds. Data taken from *Youngstown Vindicator*, 10 October 1959.

44. Ibid. In the months of August and September new welfare cases totaled 2,558 and 2,772 respectively.

45. Ibid., 8 November 1959.

46. Ibid. During the 1959 strike YS&T employee Boyd Ware "went on welfare and had four kids" (Ware interview).

47. John Occhipinti, interview by author, tape recording, Struthers, Ohio, 10 June 1993.


## Chapter 5: How to Steal a Wheelbarrow

1. While roughly 70 percent of the workers interviewed started prior to the end of World War II, only 4 percent of them could identify Gus Hall. Only three workers could identify him as a labor organizer and Communist.

2. George Papalko said that he knew of the *Daily Worker* but never read it, and both he and John Pallay agreed that the "socialists were important to organizing the union," but, significantly, that "they were out of [the workers'] league." A follow up question revealed that, "out of league" meant that workers just wanted a union, while the socialists wanted to get rid of the "free enterprise" system. Papalko, interview by author, tape recording, Youngstown, Ohio,

2 June 1993; and Pallay, interview by author, tape recording, Youngstown, Ohio, 2 June 1993.

3. Richard Oestreicher, "Separate Tribes? Working-Class and Women's History," *Reviews in American History* 19 (1991), 228–231.

4. James Lewis Baughman, "The Little Steel Strike in Three Ohio Communities" (master's thesis, Columbia University, N.Y., 1975).

5. There were of course exceptions. One such exception was Tony Nocera, who helped to organize the Republic plant in the 1930s and experienced the strike in 1937. Of the workers I spoke to, Nocera had the most knowledge about unionism in Youngstown. He had read numerous books on labor and could recite the history of events leading up to the successful recognition of SWOC in 1942.

6. Zygmunt Bauman, *Memories of Class* (London: Routledge and Kegan Paul, 1983), 13.

7. Joe Ryzner, interview by author, tape recording, Liberty Township, Ohio, 3 June 1993.

8. John McGarry, interview by author, tape recording, Struthers, Ohio, 23 June 1993.

9. Quoted from a letter from Francis Gillin, vice president of Local Union 1418 to International president David McDonald, dated 7 May 1947. Material in Historical Collection and Labor Archives, United Steelworkers of America Records, 1947, Patte Library, Pennsylvania State University, College Park, Pa., Box 1, F-7.

10. Baughman, "The Little Steel Strike."

11. *Brier Hill Unionist* (Local 1418 of YS&T), March 1975, and "The Wildcat Over Tony's Death," in *We Are the Union: The Story of Ed Mann*, ed. Alice Lynd and Staughton Lynd, distributed by Solidarity USA, 38.

12. Brad Ramsbottom, interview by author, tape recording, Struthers, Ohio, 12 June 1993.

13. While I contend there are important radical implications for the distribution of shop floor power and class consciousness in the use of the grievance system, it is undeniable that the purpose of the National Labor Relations Act was to preserve the free flow of commerce. In fact, the legislation's opening sentence admits as much: "It is the national interest of the United States to maintain full production in its economy." Excerpted from National Labor Relations Board, *A Guide to Basic Law and Procedures under the National Labor Relations Act* (Washington, D.C.: GPO, 1976), 293. The legal formalism of labor relations is superbly critiqued in James Atelson's *Values and Assumptions in American Labor Law* (Amherst, Mass.: University of Massachusetts Press, 1983) and in William Forbath's *Law and the Shaping of the American Labor Movement* (Cambridge: Harvard University Press, 1991).

14. Merlin Luce, interview by author, tape recording, Youngstown, Ohio, 11 June 1993.

15. Particular arbitration cases included four decisions from the *Office of the Umpire, Republic Steel Corporation, Youngstown District and United Steelworkers of America, Local Union No. 1331*, case nos. 0-75-64, 0-94-64, and 0-4-64, on 29 March 1966, and 0-35-76, on 20 July 1979 (George Porrazzo's personal records).

16. Quotes are taken from the original typed letter, dated 21 January 1966, kept in Bob Dill's personal records.

17. Taken from *Office of the Umpire, Republic Steel Corporation, Youngstown District and United Steelworkers of America, Local Union No. 1331,* case no. 0-66-65, on 29 March 1966 (George Porrazzo's personal records).

18. Summary of Local 1331 Grievances, 1980–1985, in George Porrazzo's personal records. While the above incidents occurred outside the time frame of this study they, nonetheless, present a picture of the scope of group grievances that workers likely constructed.

19. Anthony Delisio described a "fence" between workers and bosses that should never be crossed. Bob Dill called it a "line" that a worker had to respect or he would personally, as he said, "take him to my house or up the union hall and ball his ass out, tell him I was tired of his bullshit and that he either cooperate with the workers or he quit the damn job" (Delisio, interview by author, tape recording, Struthers, Ohio, 3 June 1993; and Dill, interview by author, tape recording, Struthers, Ohio, 13 June 1993).

20. Thomas G. Fuechtmann, *Steeples and Stacks: Religion and Steel Crisis in Youngstown,* (Cambridge: Cambridge University Press, 1989), 43.

21. Ibid.

22. Lynd points out that ABC Television did a documentary on the closings and referred to 19 September 1977 as "Black Monday." See Staughton Lynd, *The Fight against Shutdowns: Youngstown's Steel Mill Closings* (San Pedro: Singlejack, 1983), 21.

23. George Bodnar, interview by author, tape recording, Struthers, Ohio, 16 June 1993; and Bill Calabrette, interview by author, tape recording, Struthers, Ohio, 4 June 1993. Calabrette compared what the company did to irresponsible home ownership: "It was like your home. You have to keep it up. Put on new carpet and repair the roof. They just shut it down and went someplace else." Jim DeChellis felt that the mills closed because "Lykes wasn't paying the bills." He blamed Lykes for the shutdown and was adamant in pointing out that "it wasn't YS&T people who were responsible" (DeChellis, interview by author, tape recording, Struthers, Ohio, 3 June 1993).

24. Paul Dubos though that the "old" YS&T should have seen the need to modernize earlier in the century and been better prepared for the competition generated by foreign imports (Dubos, interview by author, tape recording, Campbell, Ohio, 8 July 1993). Many workers also cited imported steel, steel substitutes, environmental regulations, and government indifference as reasons for the shutdowns. Despite the array of opinions, however, every worker strongly believed that the "new" YS&T was bleeding the company.

25. I heard the term first used by my father to describe repair work. While my father worked at Republic Steel, worker complaints about the lack of company investments were typical in other mills. Dad (and other workers) was angry at company waste. He said, "They didn't care about modernizing. We were doing wire and burlap for years because Republic wouldn't invest. You had to scrounge around to find a part" (Robert Bruno, interview by author, tape recording, Struthers, Ohio, 30 August 1993). Brad Ramsbottom said, "YS&T was taking the money and paying the stockholders and doing other things with the profits instead of investing in things the mill needed. Every time you needed something you'd hear we can't afford this, we can't buy that. It was ridiculous. You didn't have the parts you needed to do the job right and you

had to bastardize" (Ramsbottom, interview). The Ohio Public Interest Campaign confirmed workers experiences by issuing a report, which among other things stated that "Lykes failed to modernize, investing an average of only $27 million annually between 1970 and 1973." See Lynd, *The Fight against Shutdowns*, 24.

26. Luce, interview.

27. It is interesting to note that Bob Dill was one of five union representatives present at the meeting with Lykes-YS&T officers when word of the shutdowns was given (Dill, interview).

28. Lykes purchased Coastal Plains Life Insurance (1969), half of a steamship company jointly owned by Lykes and W. R. Grace (1971), Ramseyer and Miller, Inc. (1973), and Great Western Steel Company (1975). See Lynd, *The Fight against Shutdowns*, 24–25.

29. See Fuechtmann, *Steeples and Stacks*, 45.

30. *Struthers (Ohio) Journal*, 28 July 1949, 11 August 1955 and 7 July 1961.

31. Ibid., 4 June 1937.

32. Material taken from *Statement of the Youngstown Sheet and Tube Company before a Panel of the National War Labor Board*, 20 May 1942 (on file in the Youngstown Public Library, reference section). Also, see the *Struthers Journal*, 18 June 1937.

33. Walter Galenson, *The CIO Challenge to the AFL: A History of the American Labor Movement, 1935–1941* (Cambridge: Harvard University Press, 1960), 109.

34. *Struthers Journal*, 14 July 1949.

35. The *Youngstown Vindicator* was quoted in Mark Reutter, *Sparrows Point: Making Steel—The Rise and Ruin of American Industrial Might* (New York: Summit Books, 1988), 192–193.

36. Fuechtmann, *Steeples and Stacks*, 14. Mark Reutter claims that over $2.25 million was spent in court fees (Reutter, *Sparrows Point*, 199).

37. Paul Tiffany, *The Decline of American Steel: How Management, Labor, and Government Went Wrong* (New York: Oxford University Press, 1998), 155–156.

38. The company bought Perault Fibercast Corporation of Tulsa, Oklahoma, to supplement its steel pipe division (*Youngstown Vindicator*, 15 August 1953).

39. For a chronology of events see Lynd, *The Fight against Shutdowns*, 6–9. John Hoerr's *And the Wolf Finally Came*, 480, includes the story of LTV's bankruptcy.

40. Albert Campbell, interview by author, tape recording, Youngstown, Ohio, 17 June 1993.

41. Calabrette, interview.

42. Bruno, interview.

43. Russ Baxter, interview by author, tape recording, Campbell, Ohio, 13 June 1993.

44. "Ye Gossip Column," *Ohio Works Organizer* (USWA Local 1330), November 1946.

45. Joe Flora was a PRO of the Month winner for turning a switch off, preventing potential serious injury to co-workers (Joe Flora, interview by author, tape recording, Struthers, Ohio, 10 June 1993). In 1973 the PRO of the Year

prize was a new Ford Pinto Wagon and tickets to a Pittsburgh Steelers professional football game (*Struthers Journal*, 30 August 1973).

46. ORP winners were usually chosen for abusive behavior and lack of concern for worker safety. But my personal favorite was the foreman selected for
"following workers to the bathroom" (*Brier Hill Unionist*, November 1975 and
February 1976).

47. *Brier Hill Unionist*, February 1976.

48. Antigio Mosconi believed that because the company did not respect the
workers "the big shots pressured the boss" to make more money at the expense
of the workers (Antigio Mosconi, interview by author, tape recording,
Youngstown, Ohio, 19 June 1993).

49. John Zumrick, interview by author, tape recording, Youngstown, Ohio, 4
June 1993.

50. *Brier Hill Unionist*, May 1975.

51. Delisio, interview.

52. Delisio, interview.

53. McGarry, interview. Workers welcomed into the "family" those bosses
that "would dirty their hands" (John Varga, interview by author, tape recording,
Struthers, Ohio, 2 June 1993).

54. Varga, interview.

55. Jim Morris, interview by author, tape recording, McDonald, Ohio, 30
June 1993; and John Occhipinti, interview by author, tape recording, Struthers,
Ohio, 10 June 1993.

56. Charley Petrunak, interview by author, tape recording, Struthers, Ohio, 3
June 1993; and Occhipinti, interview.

57. James Scott convincingly argues that subordinate groups will disguise their
true feelings about the ruling class and the dominant social relations by enacting
and disguising a public performance. See James C. Scott, *Domination and the Arts
of Resistance: Hidden Transcripts* (New Haven: Yale University Press, 1990).

58. Bruno, interview.

59. The difference can be seen by the following hypothetical comparison
based on a twenty-cent-per-piece bonus, a threshold of seven hundred pieces, a
ten-dollar hourly wage rate, and an eight-hour shift:

|                       | *Production by the Rules* |
| --------------------- | ------------------------- |
| Units                 | 900                       |
| Incentive             | 200                       |
| Hours delayed pay     | 1                         |
| Bonus earnings        | $50                       |
|                       | *Production by Inspector* |
| Units                 | 1000                      |
| Incentive             | 300                       |
| Hours delayed pay     | 2                         |
| Bonus earnings        | $80                       |

The practice of "cheating" on delays was told to me by my father (Bruno, interview).

60. John McGarry admitted that while a lot of guys would not take anything out of the plant, many would be "working on . . . individual jobs and not what [they] were supposed to do." But to the workers this was something to be respected. "If a guy said he was working on a government job you left him alone" (McGarry, interview).

61. Flora, interview.

62. Numerous interviewees told crazy stories of workers, who while trying to smuggle out everything that "wasn't tied down," stumbled about and dropped their ill-gotten goods right there at the foot of the plant guard.

63. Campbell, interview.

64. This story was told to me three separate times by workers who believed it was true. The importance of folktales, myths, and legends to the formation and power of cultures and oppressed groups is detailed in Charles Joyner's *Down by the Riverside: A South Carolina Slave Community* (Urbana: University of Illinois Press, 1984) and Scott's *Domination and the Arts of Resistance*.

65. Delisio, interview.

66. Morris, interview.

67. The "war of movement" was Antonio Gramsci's phrase for a direct forceful confrontation between labor and capital and was contrasted with a more ideological "war of position." See Joseph V. Femia, *Gramsci's Political Thought: Hegemony, Consciousness, and the Revolutionary Process* (Oxford: Clarendon Press, 1987).

## Chapter 6: A Vote for a Steelworker Is a Vote for Yourself

1. Data taken from the *Youngstown City Directories*, 1950–1985.

2. In the time period studied, there were a total of eight USWA council members who served terms exceeding three years. Mike Petruska and Walter Zalusky led the pack with eight terms, and Joe Vlosich and J. L. Williams followed with seven (*Youngstown City Directory*, 1950–1985, and Struthers City Council Meeting Minutes, 1950–1980).

3. The councilman was First Ward representative and YS&T employee William Murphy (Struthers Council Minutes, 7 May 1969).

4. John Barbero quoted in Alice Lynd and Staughton Lynd, eds., *Rank-and-File: Personal Histories by Working-Class Organizers* (New York: Monthly Review Press, 1988), 267–268.

5. Barbero, quoted in Lynd and Lynd, *Rank-and-File*, 268–269.

6. Quote taken from an early 1930s flyer announcing a mass meeting at the "Youngstown District Headquarters, 266 E. Federal St." (From a collection of Communist Party papers at the North American Labor History Conference, Wayne State University, Detroit, Mich., 1993.)

7. Statement of the Steel District, Ohio Communist Party, from the Historical Collection and Labor Archives, United Steelworkers of America Records (hereafter cited as USWA Records), 1953, Patte Library, Pennsylvania State University, College Park, Pa., Box 9, F-49.

8. Bert Cochran, *Labor and Communism: The Conflict That Shaped American Unions* (Princeton: Princeton University Press, 1977), 266–267.

9. Statement of the Ohio Communist Party, USWA Records, 1953, Box 9, F-49.

10. Quoted from an advertisement titled "Unionism—Not Sensationalism!" in the *Youngstown Vindicator,* 9 February 1953, from USWA Records, 1953, Box 9, F-49. It is worth noting that the James Griffin-Danny Thomas election was a well-publicized, vicious, internal union feud. Both candidates charged each other with unsavory practices, including connections to the "Rackets," "bribes," and "death threats" (material from USWA Records, 1953, Box 9, F-49). In the end, Griffin was reelected by a large margin.

11. Cochran, *Labor and Communism,* 154.

12. The *Struthers (Ohio) Journal,* 6 December 1946, 10 March 1949, 17 March 1949, 31 March 1949, 7 April 1949, 21 April 1949, 28 April 1949, 3 November 1949, 22 October 1953, and 30 April 1959.

13. In a political advertisement, Victor Vasvari also stressed his involvement with the "CIO since 1936," and other office holders reminded voters that "Labor Should Support Labor's Friends" (*Struthers Journal,* 17 and 31 March 1949).

14. *Struthers Journal,* 3 November 1949.

15. Richard Hamilton, *Class and Politics in the United States* (New York: John Wiley, 1972), 191.

16. Sidney Verba, Norman H. Nie, and John R. Petrocik, *The Changing American Voter* (Cambridge: Harvard University Press, 1979), 29.

17. Fay Calkins, *The CIO and the Democratic Party* (Chicago: University of Chicago Press, 1952), 33.

18. James Caldwell Foster, *The Union Politic: The CIO PAC* (Columbia, Mo.: University of Missouri Press, 1975), 209.

19. Foster, *The Union Politic,* 215.

20. *Struthers Journal,* 10 March 1949. During the 1950 political season the National CIO legislative platform included the following items: (1) fair employment practice bill, (2) outlawing the poll tax, (3) antilynching bill, (4) extension of rent control, (5) housing program, (6) minimum wage of one dollar an hour, (7) full employment, (8) national health insurance, (9) federal aid to education, (10) abolish congressional seniority system, (11) direct and open primaries. Taken from Calkins, *The CIO and the Democratic Party,* 156.

21. *Struthers Journal,* 6 October 1949.

22. An account of one of the senator's visits was told to me by James Rich, who claimed that YS&T workers left their jobs and massed near the gate entrance to yell profanities. When the senator moved out of the plant, the company fired the elected union members who had "instigated" a wildcat strike. In response, workers staged a more extensive shutdown until their leaders were put back to work (James Rich, interviewed by author, tape recording, Campbell, Ohio, 9 August 1993).

23. Survey responses taken from Gallup Poll dated November 1949, as reported in Verba, Nie, and Petrocik, *Changing American Voter,* 100.

24. A sample of issues taken from the 1960s that generated strong organized lobbying efforts includes: (1) reduction of work week, (2) hours of pay based on forty hours, (3) government medical care, (4) government support for education, (5) public housing investment, (6) rise in the minimum wage, and (7) lifting of restrictions on strike activity. From a variety of national polls union members

showed greater support on all of these items than did the general public (Dan C. Heldman and Deborah L. Knight, *Unions and Lobbying: The Representation Function*, [Arlington, Va.: Foundation for The Advancement of the Public Trust, 1980]).

25. Opinion data taken from Harris and Gallup Polls as reported in Derek Bok and John T. Dunlop, *Labor and the American Community* (New York: Touchstone Books, 1970), appendix B, 58–63.

26. Hamilton, *Class and Politics*, 206.

27. This is to be contrasted with the conventional view of labor as just another constituent socioeconomic interest group within the Democratic Party. Perhaps the most critical way to view this is to see it as co-optation and selling out (Such is Mike Davis's position in "The Barren Marriage of American Labor," *New Left Review* 124 [November/December 1980], 43–84). A work which does not diverge from a "constituent" characterization but does enlarge the role and power of organized labor is David J. Greenstone, *Labor in American Politics* (Chicago: University of Chicago Press, 1974).

28. One of the most common complaints registered by citizens was the noxious gas smells that permeated their homes (Struthers Council Minutes, 10 October 1964 and 2 February 1966). Another resident also bitterly pointed out, "In order to get the porch clean I had to use Clorox" (Struthers Council Minutes, 3 February 1960).

29. Struthers Council Minutes, 6 April 1960.

30. Quote is from Councilman Tom Vasvari, taken from the Struthers Council Minutes, 7 February 1968.

31. Struthers Council Minutes, 7 February 1968. Statement made by Councilman Yurko, who was a member of the Smoke Abatement Committee. Another member of that committee, William Murphy, even softened his stance on the problem by rewording the language in a proposed ordinance because the council "was afraid the Sheet and Tube was going to move and [he would] feel like the culprit" (Struthers Council Minutes, 7 February 1968).

32. Quote is from Dick Thompson, program director of local radio station WHOT in the *Warren-Youngstown Business Journal* (May 1993), 23.

33. At one meeting, Murphy provided visual and written records of at least seventy-five damaged homes (Struthers Council Minutes, 19 February 1964). In addition, ward residents attended council meetings to bitterly complain about the damage done to their homes. In fact, it seems that the most common questions asked by people in attendance was "Who or what is responsible for the black rain?" (Struthers Council Minutes, 3 April 1963).

34. Struthers Council Minutes, 7 September 1966.

35. Struthers Council Minutes, 20 November 1968.

36. *Struthers Journal*, 17 March 1949.

37. In 1959, Vasvari collected 2,619 votes to finish above two other USWA, at-large candidates (Michael Petruska and Walter Zalusky). All three, however, were elected to office (*Struthers Journal*, 30 April 1959).

38. He was not always so temperate. In response to a letter sent to the council from YS&T about the pollution problem, Vasvari angrily stated that it "amounted to nothing more than a shovel of words" (Struthers Council Minutes, 3 May 1962).

39. Struthers Council Minutes, 20 November 1968.

40. Struthers Council Minutes, 2 November 1966. In 1946, Katula himself was a councilman in the city of Campbell (*Struthers Journal*, 6 December 1946).

41. The letter was from Joseph Clark (Struthers Council Minutes, 2 November 1966).

42. This letter was from Thomas Finn of Local Union 1418 (Struthers Council Minutes, 2 November 1966).

43. Struthers Council Minutes, 2 November 1966.

44. An exchange between both men focused on work schedules and who was and was not paid when they attended council meetings (Struthers Council Minutes, 2 November 1966). Vasvari obviously felt haunted by the question of meeting attendance because in 1972 he read into the record the absentee records of all members of the council. While Vasvari's figures were meant to exonerate him from any claims of irresponsibility, they reveal that steelworking members, despite sometimes working three shifts, had the best attendance. The members with the fewest missed meetings were all steelworkers: Vasvari (2), Vlosich (5), and Chapella (11). Struthers Council Minutes, 6 December 1972.

45. The issue generated a good deal of political manipulation and intrigue. The original ordinance was passed without a public hearing in 1951 but due to a technicality could not become law under "emergency rules." The CIO then circulated petitions to place the issue on the ballot. Despite city efforts to squelch the citizen petition drive, the referendum did occur and was badly trashed by a three to one count. But the next elected council immediately moved to re-pass the ordinance ("Police State In Struthers," *Ohio Works Organizer* [USWA Local 1330], 7 March, 1952).

46. Struthers Council Minutes, 4 April 1953.

47. Struthers Council Minutes, 28 April 1952.

48. Struthers Council Minutes, 23 September 1959.

49. Struthers Council Minutes, 20 May 1953 and 5 September 1962.

50. Struthers Council Minutes, 19 July 1950 and 6 August 1958.

51. Struthers Council Minutes, 16 September 1961 and 3 May 1962.

52. The quote came from nonsteelworking councilman Daniel Yurko (Struthers Council Minutes, 20 October 1971).

53. Struthers Council Minutes, 5 January 1972 and 6 October 1971.

54. Struthers Council Minutes, 15 September 1971.

55. Struthers Council Minutes, 20 October 1971.

56. Struthers City Council Resolution 4754, 17 April 1963. The council did the same thing ten years earlier when YS&T president Frank Purnell retired. This time they praised the company official for being "a great pioneer who enabled Struthers to be classed as the steel center of the United States" (Struthers Council Minutes, 1 April 1953).

57. Struthers Council Minutes, 18 June 1958. It was not a little ironic that when the mills collapsed one of the most stark physical signs of an ill wind blowing was the utter decay that befell Mauthe Park.

58. Imports in 1972 failed a very modest 3.4 percent from 1971's record amount of 18.3 million tons (William T. Hogan, *Steel In Crisis* [Pittsburgh: Steel Communities Coalition, November 1977], 8–9).

59. Struthers Council Minutes, 1 March 1972.

60. Statement made by Local Union 1418 member Larry Barber (Struthers Council Minutes, 21 June 1972). The entire resolution read, "A Resolution Protesting and Objecting to the Use of Foreign Steel by American Companies" (Struthers City Council, Resolution 6451, 15 March 1972).

61. Auditor John Kovach was the most vocal city government proponent of fiscal austerity (Struthers Council Minutes, 19 May 1971).

62. Struthers Council Minutes, 20 December 1972.

63. The phrase was used a couple of times by Auditor Kovach (Struthers Council Minutes, 17 January 1973).

64. Struthers Council Minutes, 5 March 1969.

65. Struthers Council Minutes, 5 March 1969.

66. The Struthers City Council knew of YS&T-Lykes's plan to invest over $200 million in plant expansion, "but not a dime of it within the city limits" (quote from city solicitor, Struthers Council Minutes, 11 November 1973).

67. The characterization was made by Councilman Barber in reference to what would happen if the council did not act to attract new industry (Struthers Council Minutes, 5 March 1969).

68. Vlosich proposed a National Steel Policy, and the city received part of a $2.5 million aide package (Struthers Council Minutes, 19 October 1977 and 7 December 1977).

69. Struthers Council Minutes, 21 September 1977.

70. I find it incomprehensible that a major work like Verba, Nie, and Petrocik's *The Changing American Voter* could go on analyzing public opinion and political choices for 430 pages, including the index, and never once mention the word "class."

## Chapter 7: Youngstown, Once Famous for Steel, Lost Its Name

1. Robert Bruno, "Remembering Black Monday: Anatomy of a Shutdown" *Workplace Democracy* 58 (Spring 1988), 7–9. Also see "Mayor's Report," Struthers City Council Meeting Minutes, no. 4806, 21 September 1977.

2. For an outstanding account of how a coalition of community members, local unions, and activists fought to prevent an economic holocaust see Staughton Lynd, *The Fight against Shutdowns: Youngstown's Steel Mill Closings* (San Pedro: Singlejack, 1983). I will be relying primarily on Lynd's detailed reporting of the Save Our Valley Campaign throughout this section.

3. Lynd reports that as the first group of terminated YS&T workers left the Campbell Works and "crossed the Mahoning River on the footbridge which led to the clockhouse many [workers] threw into the river the hard hats and metatarsal shoes of their trade" (quoted in Lynd, *The Fight against Shutdowns*, 22).

4. Gerald Dickey, the editor of the *Brier Hill Unionist* (Local 1418 of YS &T), wrote an editorial, entitled "A Possible Alternative—Employee Ownership." Lynd quotes from the editorial in *The Fight against Shutdowns*, 29–30.

5. Lynd, *The Fight against Shutdowns*, 32–40.

6. George Bodnar worked in the open hearth at Brier Hill. He seemed to exemplify the attitude of most workers when I asked him, "Did you get involved

in the effort out of Brier Hill to buy the mill?" He replied, "No, I never did. But other workers bought plants, so I don't see why we couldn't have here" (Bodnar, interview by author, tape recording, Campbell, Ohio, 16 June 1993).

7. Lynd, *The Fight against Shutdowns,* 151–159.

8. Lynd, *The Fight against Shutdowns,* 49–62.

9. The *Brier Hill Unionist* reported that 2,000 accounts, worth approximately two million dollars had been taken out (April–May 1978).

10. See Lynd, *The Fight against Shutdowns,* 123–125.

11. For a description of key workers involved in the Save Our Valley Campaign see Lynd, *The Fight against Shutdowns,* 9–11.

12. Wage increases, however, had a dual history. Prior to the "no-strike" period (1961–1970), real hourly earnings rose 3.76 percent a year. But with the signing of the Experimental Negotiating Agreement (ENA), wages rose only 0.44 percent a year (Lynd, *The Fight against Shutdowns,* 54).

13. Mark McColloch, "Consolidating Industrial Citizenship: The USWA at War and Peace, 1939–46," in *Forging a Union of Steel: Philip Murray, SWOC, and the United Steelworkers,* ed. Paul Clark, Peter Gottlieb, and Donald Kennedy (Ithaca, N.Y.: ILR Press, 1987), 45–86.

14. Albert Campbell, interview by author, tape recording, Youngstown, Ohio, 17 June 1993; Jim and Josephine DeChellis, interview by author, tape recording, Struthers, Ohio, 3 June 1993; and Chris Cullen, interview by author, tape recording, Youngstown, Ohio, 12 July 1993.

15. Tony Modarelli, interview by author, tape recording, Struthers, Ohio, 4 June 1993.

16. George Porrazzo, interview by author, tape recording, Struthers, Ohio, 12 June 1993.

17. Quoted from Ed Mann, "From the President's Desk . . ." *Brier Hill Unionist* (May 1975). Youngstown's Local 1462 and 1330 had progressive left leadership.

18. John Pallay, interview by author, tape recording, Youngstown, Ohio, 2 June 1993; and George Papalko, interview by author, tape recording, Youngstown, Ohio, 2 June 1993.

19. Also see, Ed Mann, videotape recording, Ohio Historical Society, Youngstown Center of Industry and Labor, Youngstown, Ohio, 16 April 1991.

20. John Hoerr makes the argument that the ENA gave steelworkers the better end of the deal. He also contends that while there were bitter internal fights over the ENA, most rank-and-filers were willing to trade the right to strike for an assurance of economic prosperity. See John P. Hoerr, *And the Wolf Finally Came: The Decline of the American Steel Industry* (Pittsburgh: University of Pittsburgh Press, 1988), 109–121.

21. Lynd, *The Fight against Shutdowns,* 53.

22. John Occhipinti, interview by author, tape recording, Struthers, Ohio, 10 June 1993; Sam Shapiro, interview by author, tape recording, Youngstown, Ohio, 10 August 1993.

23. John Druzoff agreed that without the strike, workers got bogged down in "formal procedures that took too long" (Druzoff, interview by author, tape recording, Austintown, Ohio, 12 July 1993).

24. While most workers had little detailed knowledge of their company's senior managers, they all assumed that presidents and vice presidents made a great deal of money and "lived high off the hog" (DeChellis, interview). Workers were very generous in believing that supervisors should make good money, and some even believed that production workers should earn less than their bosses; however, workers condemned the vast distance between what they earned and what managers walked away with.

25. Joseph Lykes, $218,910; Frank Nemec, $218,910; Jennings Lamberth, $159,207; Thomas Clearly, $143,287; Randolph Rieder, $143,287; George Kimmel, $119,408. Salary figures taken from the *Brier Hill Unionist,* February–March 1977.

26. Porrazzo, interview.

27. John Varga pointed out that in the plant, nationality determined job placement, but in the corporate boardroom shrewd, intelligent men ran the operations. He had a near devotion for ex-USS president Edgar Speer. When I asked him, "Who was responsible for the value created in production?" he responded, "Edgar Speer and the guys doing the work for him" (Varga, interview by author, tape recording, Struthers, Ohio, 2 June 1993). Many men also believed that formal education separated them from company executives. Russ Baxter felt that in order to be an executive or company owner "you had to have some brains and have gone to college" (Baxter, interview by author, tape recording, Campbell, Ohio, 13 June 1993). In most cases, however, workers credited the social advantages of bosses, executives, and stockholders more to bloodline than to hard work. As Bill Calabrette put it, "Money stayed with money; it came down through the family" (Calabrette, interview by author, tape recording, Struthers, Ohio, 4 June 1993).

28. Jim Visingardi, interview by author, tape recording, Struthers, Ohio, 17 June 1993.

29. Cayetano Caban, interview by author, tape recording, Struthers, Ohio, 17 June 1993.

30. John McGarry, interview by author, tape recording, Struthers, Ohio, 23 June 1993.

31. Theories of class convergence are discussed by Herbert Gutman in "Historical Consciousness in Contemporary America," in his *Power and Culture: Essays on the Class Structure,* ed. Ira Berlin (New York: Pantheon Books, 1987), 408–410.

32. David G. Stratman, *We Can Change the World: The Real Meaning of Everyday Life* (Boston: New Democracy Books, 1993), 259.

33. Robert Bruno, "When the Steel Mills Closed," *U.S. News and World Report* 105, no. 7 (August 1988), 8.

# Bibliography

## Books

Atelson, James. *Values and Assumptions in American Labor Law.* Amherst, Mass.: University of Massachusetts Press, 1983.

Bauman, Zygmunt. *Memories of Class.* London: Routledge and Kegan Paul, 1983.

Bell, Daniel. *The End of Ideology.* Glencoe, Ill.: Free Press, 1960.

Bok, Derek, and John T. Dunlop. *Labor and the American Community.* New York: Touchstone Books, 1970.

Buss, Terry F., and F. Stevens Redburn. *Reemployment after a Shutdown: The Youngstown Steel Mill Closings, 1977–1985.* Youngstown, Ohio: Center for Urban Studies, Youngstown State University, 1986.

Calkins, Fay. *The CIO and the Democratic Party.* Chicago: University of Chicago Press, 1952.

Cochran, Bert. *Labor and Communism: The Conflict That Shaped American Unions.* Princeton: Princeton University Press, 1977.

Dahrendorf, Ralf. *Class and Class Conflict in Industrial Society.* Stanford: Stanford University Press, 1959.

Davis, Mike. *Prisoners of the American Dream.* London: Verso Press, 1986.

Drucker, Peter. *The New Society: The Anatomy of the Industrial Order.* New York: Harper, 1950.

Fantasia, Rick. *Cultures of Solidarity: Consciousness, Action, and Contemporary American Workers.* California: University of California Press, 1988.

Femia, Joseph V. *Gramsci's Political Thought: Hegemony, Consciousness, and the Revolutionary Process.* Oxford: Clarendon Press, 1987.

Forbath, William. *Law and the Shaping of the American Labor Movement.*
     Cambridge: Harvard University Press, 1991.
Foster, James Caldwell. *The Union Politic: The CIO, PAC.* Columbia, Mo.:
     University of Missouri Press, 1975.
Frisch, Michael, and Milton Rogovin. *Portraits In Steel.* Ithaca, N.Y.:
     Cornell University Press, 1993.
Fuechtmann, Thomas G. *Steeples and Stacks: Religion and Steel Crisis in
     Youngstown.* Cambridge: Cambridge University Press, 1989.
Galbraith, John Kenneth. *The Affluent Society.* Boston: Houghton Mifflin,
     1958.
Galenson, Walter. *The CIO Challenge to the AFL: A History of the American
     Labor Movement, 1935–1941.* Cambridge: Harvard University Press, 1960.
Goldthorpe, John, David Lockwood, Frank Bechhoffer, and Gennifer Platt.
     *The Affluent Worker in the Class Structure.* Cambridge: Cambridge Uni-
     versity Press, 1969.
Greenstone, David J. *Labor in American Politics.* Chicago: University of
     Chicago Press, 1974.
Gutman, Herbert. *Power and Culture: Essays on the American Working
     Class.* Edited by Ira Berlin. New York: Pantheon Books, 1987.
Halle, David. *America's Working Man.* Chicago: University of Chicago
     Press, 1984.
Hamilton, Richard. *Class and Politics in the United States.* New York: John
     Wiley, 1972.
Heldman, Dan C., and Deborah L. Knight. *Unions and Lobbying: The
     Representation Function.* Arlington, Va.: Foundation for the
     Advancement of the Public Trust, 1980.
Hodgson, Godfrey. *America in Our Time.* New York: Vintage Books, 1976.
Hoerr, John P. *And the Wolf Finally Came: The Decline of the American
     Steel Industry.* Pittsburgh: University of Pittsburgh Press, 1988.
Joyner, Charles. *Down by the Riverside: A South Carolina Slave
     Community.* Urbana: University of Illinois Press, 1984.
Katznelson, Ira. "Working-Class Formation: Constructing Cases and
     Comparisons." In *Working-Class Formation: Nineteenth-Century Pat-
     terns in Western Europe and the United States.* Edited by Ira
     Katznelson and Aristide R. Zolberg. Princeton: Princeton University
     Press, 1986.
——. *Marxism and the City.* New York: Oxford University Press, 1992.
Kerr, Clark, John Dunlop, Frederick H. Arbison, and Charles M. Myers.
     *Industrialism and Industrial Man.* London: Oxford University Press, 1962.
Kolko, Gabriel. *Wealth and Power in America: An Analysis of Social Class
     and Income Distribution.* New York: Praeger Press, 1962.
Lane, Robert. *Political Ideology: Why the American Common Man Believes
     What He Does.* Glencoe, Ill.: Free Press, 1972.
Lebowitz, Michael. *Beyond Capital: Marx's Political Economy of the Work-
     ing Class.* New York: St. Martin's Press, 1992.

Levison, Andrew. *Working-Class Majority*. New York: Penguin Books, 1975.

Lipset, Seymour Martin. *Political Man: The Social Basis of Politics*. New York: Anchor Books, 1963.

Lockwood, David. "Sources of Variation in Working-Class Images of Society." In *Classes, Power, and Conflict*. Edited by Anthony Giddens and David Held. Berkeley: University of California Press, 1982.

Lynd, Alice, and Staughton Lynd. *Rank-and-File: Personal Histories by Working-Class Organizers*. New York: Monthly Review Press, 1988.

Lynd, Staughton. *The Fight against Shutdowns: Youngstown's Steel Mill Closings*. San Pedro, Calif.: Singlejack, 1983.

McColloch, Mark. "Consolidating Industrial Citizenship: The USWA at War and Peace, 1939–46." In *Forging a Union of Steel: Philip Murray, SWOC, and the United Steelworkers*. Edited by Paul Clark, Peter Gottlieb, and Donald Kennedy. Ithaca, N.Y.: ILR Press, 1987.

Morrison, Samuel Eliot, and Henry Steele Commager. *The Growth of the American Republic*. New York: Oxford University Press, 1969.

Nyden, Philip. *Steelworkers Rank and File: The Political Economy of a Union Movement*. New York: Praeger Press, 1984.

O'Toole, James, et al. *Work in America*. Cambridge: MIT Press, 1973.

Parker, Richard. *The Myth of the Middle Class: Notes on Affluence and Equality*. New York: Liveright, 1972.

Parkin, Frank. *Class Inequality and Political Order: Social Stratification in Capitalist and Communist Societies*. New York: Praeger Press, 1972.

Reutter, Mark. *Sparrows Point: Making Steel—The Rise and Ruin of American Industrial Might*. New York: Summit Books, 1988.

Riesman, Steven. *City Games: The Evolution of American Urban Society and the Rise of Sports*. Urbana: University of Illinois Press, 1989.

Schlesinger, Arthur M., Jr. *The Vital Center: The Politics of Freedom*. Boston: Houghton Mifflin, 1949.

Schumpeter, Joseph. *Imperialism and Social Class*. Translated by Heinz Norden. New York: Augustus M. Kelley, 1951.

Scott, James C. *Domination and the Arts of Resistance: Hidden Transcripts*. New Haven: Yale University Press, 1990.

Stone, Katherine. "Origins Of Job Structures in the Steel Industry." In *Labor Market Segmentation*. Edited by Richards Edwards, Michael Reich, and David Gordon. New York: D. C. Heath and Company, 1975.

Stratman, David G. *We Can Change the World: The Real Meaning of Everyday Life*. Boston: New Democracy Books, 1993.

Tiffany, Paul. *The Decline of American Steel: How Management, Labor and Government Went Wrong*. New York: Oxford University Press, 1998.

Verba, Sidney, Norman H. Nie, and John R. Petrocik. *The Changing American Voter*. Cambridge: Harvard University Press, 1979.

## Articles

Bruno, Robert. "When the Steel Mills Closed." *U.S. News and World Report*. 105, no. 7 (August 1988): 8.

——. "Remembering Black Monday: Anatomy of a Shutdown." *Workplace Democracy* 58 (Spring 1998): 7–9.

Cozen, Kathleen Neils. "Immigrants, Immigrant Neighborhoods, and the Ethnic Identity: Historical Issues." *Journal of American History* 66, no. 3. (December 1979): 603–615.

Davis, Mike. "The Barren Marriage of American Labor." *New Left Review* 124 (November/December 1980): 43–84

DuBois, Tom. "Steel: Past the Crossroads." *Labor Research Review* (Winter 1983): 3–11.

——. "The Myth of the $26-an-Hour Steelworker." *Labor Research Review* (Winter 1983): 52–53.

Foner, Eric. "Why Is There No Socialism in the United States?" *History Workshop Journal* no. 17 (1984): 57–80.

Highman, John. "The Cult of the American Consensus: Homogenizing Our History." *Commentary* 27 (February 1959): 93–100.

Kazin, Michael. "Struggling with Class Struggle: Marxism and the Search for a Synthesis of U.S. Labor History." *Labor History* 28 (1987): 497–514.

Kerr, Clark. "Industrial Conflict and Its Mediation." *American Journal of Sociology* 60, no. 3 (November 1954): 230–245.

Legget, John. "Economic Insecurity and Working-Class Consciousness." *American Sociological Review* 29 (1968): 226–234.

Metzgar, Jack. "Would Wage Concessions Help the Steel Industry?" *Labor Research Review* (Winter 1983): 26–37.

Oestreicher, Richard. "Separate Tribes? Working-Class and Women's History." *Reviews in American History* 19 (1991): 228–231.

Pred, Allan. "Production, Family, and Free-Time Projects: A Time-Geographic Perspective on the Individual and Societal Change in Nineteenth-Century U.S. Cities." *Journal of Historical Geography* 7 (January 1981): 3–36.

Reisman, Frank, and S. M. Miller. "Are Workers Middle Class?" *Dissent* 8 (Fall 1961): 507–516.

Vanneman, Reeve, and Fred C. Pamel. "The American Perception of Class and Social Status." *American Sociological Review* 42, no. 3 (June 1977): 422–437.

## Newspapers

*The Business Journal.* May 1993.

*The Militant.* 9 January 1950–10 February 1956.

*The Struthers Journal.* 4 June 1937–30 August 1973.

*The Youngstown Vindicator.* 21 November 1915–31 August 1993.

## Worker Interviews

Andrews, Kenneth. Interview by author. Tape recording. Youngstown, Ohio, 30 June 30 1993.

Bator, Steven. Interview by author. Tape recording. Austintown, Ohio, 12 July 1993.

Baxter, Russell. Interview by author. Tape recording. Campbell, Ohio, 13 June 1993.

Beatty, Elma Jones. Transcripts of interview, 28 August 1991. Ohio Historical Society Oral History Project. Youngstown Historical Center of Industry and Labor. Youngstown, Ohio.

Beck, Carl. Transcript of interview. 8 May 1974. Ohio Historical Society Oral History Project. Youngstown Historical Center of Industry and Labor. Youngstown, Ohio.

Bergman, Thomas and Lillian. Interview by author. Tape recording. Struthers, Ohio, 7 June 1993.

Bodnar, George. Interview by author. Tape recording. Struthers, Ohio, 16 June 1993.

Bruno, Robert and Lena. Interview by author. Taped recording. Struthers, Ohio, 30 August 1993.

Caban, Cayetano and Consuelo. Interview by author. Tape recording. Campbell, Ohio, 17 June 1993.

Calabrette, William and Janice. Interview by author. Tape recording. Struthers, Ohio, 4 June 1993.

Campbell, Albert. Interview by author. Tape recording. Youngstown, Ohio, 17 June 1993.

Carlini, Joseph. Interview by author. Tape recording. Struthers, Ohio, 9 June 1993.

Costello, John. Interview by author. Tape recording. Youngstown, Ohio, 29 June 1993.

Cox, Margaret. Interview by author. Tape recording. Youngstown, Ohio, 4 June 1993.

Crivelli, Mario. Interviewed by author. Tape recording. New Bedford, Pa., July 7, 1993.

Cullen, Christopher and Katherine. Interview by author. Tape recording. Youngstown, Ohio, 12 July 1993.

Davis, James. Transcript of interview, 22 April 1991. Ohio Historical Society Oral History Project. Youngstown Historical Center of Industry and Labor. Youngstown, Ohio.

DeChellis, James and Josephine. Interview by author. Tape recording. Struthers, Ohio, 3 June 1993.

Delisio, Anthony and Joanne. Interview by author. Tape recording. Struthers, Ohio, 3 June 1993.

Delquadri, Anthony and Mary. Interview by author. Tape recording. Youngstown, Ohio, 7 June 1993.

Dill, Robert. Interview by author. Tape recording. Struthers, Ohio, 13 June 1993.

Donnely, Walter. Interview by author. Tape recording. Youngstown, Ohio,
    21 June 1993.
Dubos, Paul. Interview by author. Tape recording. Campbell, Ohio, 8 July
    1993.
Dzuroff, John. Interview by author. Tape recording. Austintown, Ohio, 12
    July 1993.
Esparro, Porfirio. Interview by author. Tape recording. Campbell, Ohio, 31
    July 1993.
Flemming, Oscar. Interview by author. Tape recording. Youngstown, Ohio,
    9 July 1993.
Flora, Joseph and Eileen. Interview by author. Tape recording. Struthers,
    Ohio, 10 June 1993.
Floyd, Bert. Interview by author. Tape recording. Youngstown, Ohio, 5
    July 1993.
Floyd, William. Interview by author. Tape recording. Youngstown, Ohio, 20
    June 1993.
Frattaroli, Frank and Ann. Interview by author. Tape recording.
    Youngstown, Ohio, 8 June 1993.
Freeman, Charles. Interview by author. Tape recording. Youngstown, Ohio,
    1 July 1993.
Harp, Charles. Interview by author. Tape recording. Campbell, Ohio, 2
    July 1993.
Izzo, Augustine. Transcript of interview, 26 March 1993. Ohio Historical So-
    ciety Oral History Project. Youngstown Historical Center of Industry and
    Labor. Youngstown, Ohio.
Jackson, Clingan. Transcript of interview, 24 April 1976. Ohio Historical So-
    ciety Oral History Project. Youngstown Historical Center of Industry and
    Labor. Youngstown, Ohio.
Kerrick, Frank. Interview by author. Tape recording. Boardman, Ohio, 23
    June 1993.
Kotasek, Thomas. Interview by author. Tape recording. Struthers, Ohio, 11
    June 1993.
Luce, Merlin. Interview by author. Tape recording. Youngstown, Ohio, 11
    June 1993.
Mann, Edward. Videotape interview, 16 April 1991. Ohio Historical Society
    Oral History Project. Youngstown Historical Center of Industry and
    Labor. Youngstown, Ohio.
McGarry, John. Interview by author. Tape recording. Struthers, Ohio, 23
    June 1993.
Modarelli, Tony and Viola. Interview by author. Tape recording. Struthers,
    Ohio, 4 June 1993.
Morris, James. Interview by author. Tape recording. McDonald, Ohio, 30
    June 1993.
Mosconi, Antigio. Interviewed by author. Tape recording. Youngstown,
    Ohio, 19 June 1993.

Mullins, Arnette. Interview by author. Tape recording. Youngstown, Ohio, 31 July 1993.

Newell, Arthur and Marcia. Interview by author. Tape recording. Struthers, Ohio, 13 June 1993.

Nocera, Tony. Interview by author. Tape recording. Campbell, Ohio, 9 August 1993.

Occhipinti, John and Helen. Interview by author. Tape recording. Struthers, Ohio, 10 June 1993.

Pajatsch, Ernest. Interview by author. Tape recording. Struthers, Ohio, 21 June 1993.

Pallay, John and Grizzel. Interview by author. Tape recording. Youngstown, Ohio, 2 June 1993.

Papalko, George. Interview by author. Tape recording. Youngstown, Ohio, 2 June 1993.

Pechuta, John. Interview by author. Tape recording. Struthers, Ohio, 9 June 1993.

Pellota, Anthony. Interview by author. Tape recording. Struthers, Ohio, 11 August 1993.

Perry, Tony. Interview by author. Tape recording. Hubbard, Ohio, 31 July 1993.

Petrunak, Charles and Florence. Interview by author. Tape recording. Struthers, Ohio, 3 June 1993.

Porrazzo, George. Interview by author. Tape recording. Struthers, Ohio, 9 June 1993.

Ramirez, Ramon. Interview by author. Tape recording. Campbell, Ohio, 18 June 1993.

Ramsbottom, Bradley. Interview by author. Tape recording. Struthers, Ohio, 12 June 1993.

Rich, James and Margaret. Interview by author. Tape recording. Campbell, Ohio, 9 August 1993.

Rodriguez, Augustine. Interview by author. Tape recording. Campbell, Ohio, 21 June 1993.

Rodriguez, Roman-Rosa. Interviewed by author. Tape Recording. Youngstown, Ohio. 16 June 1993.

Ross, Mary. Interview by author. Tape recording. Youngstown, Ohio, 15 June 1993.

Rucci, Armando. Interviewed by author. Tape recording. Struthers, Ohio, 7 June 1993.

Ryzner, Joseph and Pauline. Interview by author. Tape recording. Liberty Township, Ohio, 3 June 1993.

Sanchez, Augustine and Cora. Interview by author. Tape recording. Struthers, Ohio, 29 June 1993.

Scocca, Michael. Interview by author. Tape recording. Struthers, Ohio, 14 June 1993.

Shapiro, Samuel. Interview by author. Tape recording. Youngstown, Ohio, 10 August 1993.

Standwood, Charles. Interview by author. Tape recording. New Bedford, Pa.,
    7 July 1993.
Thomas, Daniel. Transcript of interview, 1 August 1974. Ohio Historical So-
    ciety Oral History Project. Youngstown Historical Center of Industry and
    Labor. Youngstown, Ohio.
Varga, John and Helen. Interview by author. Tape recording. Struthers, Ohio,
    2 June 1993.
Vasilchek, Michael. Interview by author. Tape recording. Youngstown,
    Ohio, 6 July 1993.
Visingardi, James. Interview by author. Tape recording. Struthers, Ohio, 17
    June 1993.
Ware, Boyd. Interview by author. Tape recording. Campbell, Ohio, 3 July
    1993.
White, Thomas. Transcript of interview, 9 July 1974. Ohio Historical
    Society Oral History Project. Youngstown Historical Center of Industry
    and Labor. Youngstown, Ohio.
Wienstock, Marvin. Transcript of interview. 28 August 1991. Ohio Histori-
    cal Society Oral History Project. Youngstown Historical Center of Indus-
    try and Labor. Youngstown, Ohio.
Zumrick, John and Rose. Interview by author. Tape recording. Youngstown,
    Ohio, 4 June 1993.

## Government Records

Mahoning County Department of Deeds and Records. Mahoning County
    Courthouse. Youngstown, Ohio.
Mahoning County Tax Records. Mahoning County Courthouse. Department
    of Tax Assessor. Youngstown, Ohio.
National Labor Relations Board. *A Guide to Basic Law and Procedures
    under the National Labor Relations Act.* Washington, D. C.: Government
    Printing Office, 1976.
United States Department of Commerce, Bureau of the Census. "General
    Characteristics of the Population by Census Tracts, Youngstown, Ohio."
    In *United States Census of Population, 1960.* Washington, D.C.: Govern-
    ment Printing Office, 1961.
United States Department of Commerce, Bureau of the Census. "Housing
    Units by Census Tract, Youngstown, Ohio." In *United States Census Of
    Population, 1960.* Washington, D.C.: Government Printing Office, 1961.
United States Department of Commerce, Bureau of the Census. "Labor
    Force Characteristics of the General Population, Youngstown, Ohio." In
    *United States Census of Population, 1960.* Washington, D.C.:
    Government Printing Office, 1961.
United States Department of Commerce, Bureau of the Census. "Number
    and Distribution of the Population by Race, Sex, Median Age, and Census

Tracts, Youngstown, Ohio, 1960." In *United States Census of Population, 1960*. Washington, D.C.: Government Printing Office, 1961.
United States Department of Commerce, Bureau of the Census. "Youngstown, Ohio Census of Foreign Born Population, 1900–1960." In *United States Census of Population, 1960*. Washington, D.C.: Government Printing Office. 1961.

## Church Records

Baptismal Register. Saint Nicholas Parish. Youngstown Diocese. 16 August 1944–26 December 1954.
Church Bulletin. Saint Nicholas Parish. Youngstown Diocese. 19–26 April and 3–10 May 1959; 31 January 1965; 25 September–16 October 1977.
*Spirit*. Saint Nicholas Parish Weekly. Youngstown Diocese. 14 March 1965.

## Local Published Sources

Cradle of Steel Program, 1948.
*50 Years of Steel*, published by YS&T in 1950.
*Impact: The Rank and File Newsletter*, July 1993.
*Industry and Commerce in Youngstown*. Youngstown Committee of the Ohio Sesquicentennial Commission, 1803–1953. Youngstown, 1953.
*100th Anniversary Program, 1776–1976*. Italian-American Club Bicentennial, Mahoning Country, Ohio, 4 July 1976.
Rosters and attendance records of St. Anthony's Italian Club in Struthers, 1950–1970.
Struthers City Council Meeting Minutes, 1950–1980.
*Struthers Echo Class Yearbook*, 1940.
*Youngstown City Directories*, 1950–1985.

## Theses

Baughman, James Lewis. "The 1937 Little Steel Strike In Three Ohio Communities." Master's thesis, Columbia University, 1975.
Beelen, George D. "The Negro in Youngstown: Growth of a Ghetto." Kent State University, 1967.

## Miscellaneous Sources

Communist Party Flyer on Youngstown. Collection of papers at the North American Labor History Conference. Wayne State University, 1993.

Delisio, Robert. "History of Loweville and the First Italian Immigrants That Arrived," 12 February 1975.

Franz-Goldman, Christine. "Socioeconomic Costs and Benefits of the Community-Worker Plan to the Youngstown-Warren SMSA." *Youngstown Demonstration Planning Project: Final Report*. Washington, D.C: National Center for Economic Alternatives. 1978.

Gottlieb, Bertram. "Methods of Deception." Industrial Department of the AFL-CIO, 1964.

Hogan, William T. *Steel In Crisis*. Pittsburgh: Steel Communities Coalition, November 1977.

*Labor Party Advocates*. Informational Newsletter.

Lynd, Alice, and Staughton Lynd, eds. *We Are the Union: The Story of Ed Mann*. Distributed by Solidarity USA.

*The Reporter*. An LTV Publication, May 1993.

## Personal Records

Campbell, Albert. Youngstown Sheet and Tube. "Coke Plant Sequence."

Delisio, Anthony. Youngstown Sheet and Tube. Medical records and correspondence with company over medical condition, 23 February 1981.

Delquadri, Anthony. Republic Steel. Savings account book (1945–1975), credit statements and bills.

Dill, Bob. Youngstown Sheet and Tube. Union election materials and personal correspondence with company (1964–1977).

Flora, Joe. Youngstown Sheet and Tube. Property deed to home, financial records (1956–1974) and tax records.

Kraynak, Tom. Youngstown Sheet and Tube employee. Notes on work schedule (1953–1972).

Porrazzo, George. Republic Steel. Union election materials, personal financial and medical records (1950–1980).

Rucci, Armando. Personal financial records (1950–1970).

Vasilchek, Mike. Youngstown Sheet and Tube. Property deed to home (16 September 1949).

## Union Records

AFL-CIO 1331 Local News. 29 October 1969.

*Brier Hill Unionist*. Paper published by Local 1462 (1973–1979).

*Campbell Open Hearth Seniority List*. Youngstown Sheet and Tube, 1970.

*Consent Decree I*. The United States Court for the Northern District of Alabama, Southern Division, Civil Action No. 74P339, 12 April 1974.

Grievance Meeting Minutes. Local Union 2163, 22 November 1966; Local
    Union 1331, 29 March 1966 and 20 July 1979; and Local Union 1331,
    1980–1985.
*Job Description and Classification Manual.* United Steelworkers of
    America, AFL-CIO and Coordinating Committee Steel Companies, 1 Jan-
    uary 1963.
*Ohio Works Organizer.* Paper published by Local 1330 (1948–1962).
*The Oldtimer.* Publication of the United Steelworkers of America.
*Record Book of Local Union 2163,* 11 December 1950–19 April 1952.
*Statement of the Youngstown Sheet and Tube Company before a Panel of
    the National War Labor Board,* 20 May 1942.
*Trade and Craft Utilization Analysis and Goals and Timetables, for Brier
    Hill and Campbell Plants.* Youngstown Sheet and Tube, 2 August 1979.
*Youngstown Sheet and Tube Company—Youngstown District, Report 1615,
    Union Lodge Report.* United Steelworkers of America, 27 January 1973.


## Libraries and Historical Collections

Historical Collection and Labor Archives, United Steelworkers of America
    Records, 1950–1960. Pattee Library. Penn State University. Box 23, F-26;
    5, F-33; 9, F-49; 30, F-21; 1, F-7. Patte Library, Pennsylvania State Univer-
    sity, College Park, Pa.
Struthers Historical Society. Struthers, Ohio.
Tamiment Institute. Bobst Library, New York University, New York, New
    York.
Youngstown Historical Center of Industry and Labor, Youngstown, Ohio.
Youngstown Public Library, Youngstown, Ohio.

# Index